ARTISTIC
PHOTOGRAPHIC
PROCESSES

ARTISTIC PHOTOGRAPHIC PROCESSES

BY SUDA HOUSE

AMPHOTO BOOKS
American Photographic Book Publishing
An imprint of Watson-Guptill Publications
1515 Broadway, New York, N.Y. 10036

All photographs not otherwise credited are by the author.

First published in New York, New York, by American Photographic Book
Publishing: an imprint of Watson-Guptill Publications, a division of
Billboard Publications, Inc., 1515 Broadway, New York, NY 10036

Library of Congress Cataloging in Publication Data

House, Suda, 1951–
  Artistic Photographic Processes.

  Bibliography
  Includes index.
  1. Photography—Special effects. 2. Photography,
Handworked. 3. Copy art. I. Title.

TR148.H68   770'.28   81-8027
ISBN 0-8174-3540-9   AACR2
ISBN 0-8174-3541-7 (pbk.)

MANUFACTURED IN THE UNITED STATES OF AMERICA

1 2 3 4 5 6 7 8 9/86 85 84 83 82 81

Dedication

This book is dedicated to my namesake and late great-grandmother, Suda Chiles Wells—a lady who didn't talk about taking chances, but did.

Acknowledgments

Although only my name appears as the author of this book, it is the sum of many people's efforts. I wish to thank the following for their help, support, and friendship beyond the call of duty:

Lou Jacobs, Jr., who approached me with the original concept for the book, for his continuing support throughout the writing process and, of course, for his timely and much-needed editing skills.

Herb Taylor, for his patience and confidence in me.

The entire Photography Department and the students of East Los Angeles College, and especially my friend Mei Valenzuela, for her expertise and moral support.

Laura Aguilar, whose fascination with all the wonders of nonsilver photography gave me new energy to continue at a crucial time.

My friend Lance Carlson, for his imagery and editorial assistance during the final stages of the project.

Dan Vandevier, who challenged me with many new ideas about nonsilver photography.

Ardon Alger and his students at Chaffey College, who believed in my concepts and continued to push their ideas and imagery long after our workshop ended.

Darryl Curran with admiration and respect, not only as a former student but as one photographer to another, and most of all as a friend.

All the artists and photographers who submitted images for inclusion.

My sisters Mary and Christy, my grandmother, my dad and mom, who knew all along I could do it.

And my special thanks to the many friends—De Ann, David, Barbara, Joan, and Bob—who listened over coffee for long hours about "the book."

CONTENTS

Introduction

This book demands that you examine critically your definition of photography. You should set aside all your previous notions about nonsilver and alternative photographic processes as well, because every idea you may have had concerning them is about to be challenged.

This is intended to be more than a how-to book. Anyone can acquire the information necessary to execute these processes, but only a few people develop a real sensitivity for them. There is great latitude for experimentation, accidents, and intuitive chance-taking. Printing camera-formed images on artist's paper or fabric is fun, and the thrill of working through formulas, solving problems with a process, and watching it develop can provide a real high.

Do not approach these processes as "just exercises" in learning new photographic techniques. Increasing your vocabulary of processes is important, but appreciating, and then correctly applying, this knowledge leads to genuine learning and personal growth. To think of the information supplied in this book as defining just another photographic technique will limit, not broaden, your understanding of photography.

Examine not only this book but also your own ideas, your images, and the ways they relate to the processes presented. Success comes with blending process and idea. It will be frustrating at times, but the results will be exciting and the rewards satisfying.

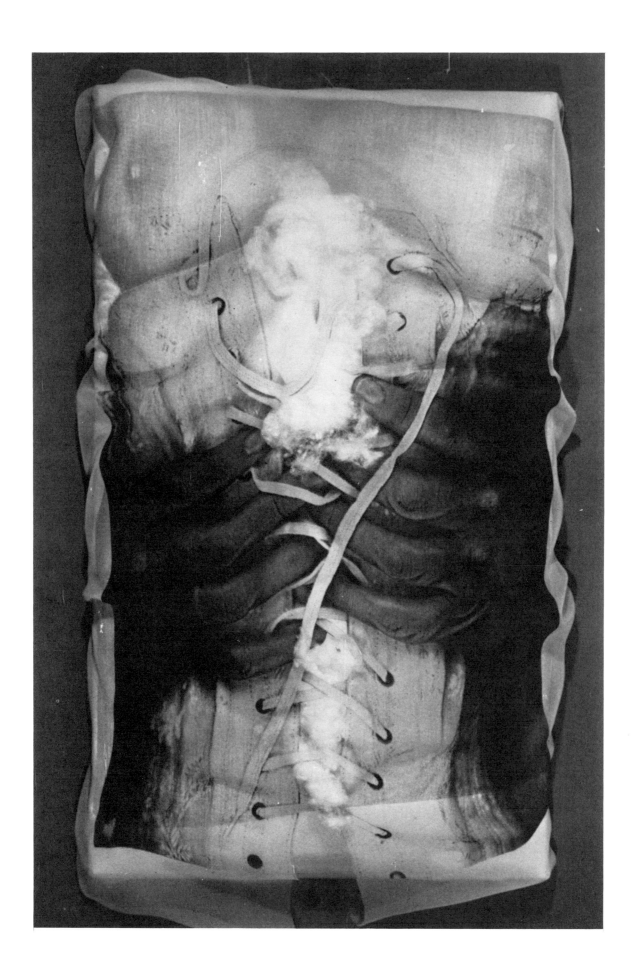

1

EXAMINING THE CREATIVE PROCESS

You Don't Know What It's Like to Have a Sex Symbol Under Your Bathrobe *is a sequence of five images, one of which is shown above. Artist Suda House created the satiric progression by transferring color Xeroxes to satin and chiffon, which was then sewn and stuffed with quilting batting.*

Students of artistic photographic processes often ask, "What is a good image to print?" or "What sort of pictures look good in blueprint or in gum print?" These are good questions, but they have no easy answers. All the images included in this book work well with the processes selected, but the photographers may not be able to pinpoint exactly *why* they select one image rather than another.

The creative process—in this case, coming up with an idea for a photographic image and then executing it—is complex; entire books deal with it. In *A Creative Approach to Controlling Photography* (Heidelberg Press, 1974) Harry Boyd writes:

> "The problem of transforming the abstract idea into a physical reality has always been the most profound challenge to confront the artist. Even for the photographer who can involve mechanical tools in the creative process, the problem of communicating the idea to the viewer is not lessened. If anything it is increased because a photographer must become the master of many procedures and devices before his medium will perform to his desires and conform to the needs of his emotions and imagination."

Where does an idea come from? How does an artist, whether painter or photographer, get an idea for an image? Does one just pop into the head like a light bulb being turned on, or is there more going on than can be imagined? I believe the latter to be the case.

An idea for a photograph—the forming of a visual image in the mind—is an accumulation of years of seeing that literally "clicks." One must be open and responsive to visual stimuli, as well as to signals from one's other senses. Recognizing something as a potential image—recording it with a camera, writing it in a journal as something to be set up later, or making a mental note of it—is the first stage in the creative process.

Selecting the medium in which such an idea is to be presented is the next stage. For many photographers the picture-taking and picture-making process involves camera, silver-imaging materials, and a darkroom. For the photographer working in nonsilver processes, the procedure is more complex. It includes not only camera, film, darkroom, and other conventional photographic tools, but also a

13

printing process, a printing surface, and the various finishing techniques.

Control is the next stage for the photographer experimenting with alternative processes. Not only must the image be perceived and photographed; it must also be translated step by step through numerous procedures and evaluated at each point to make sure the image still fits the original idea.

The continuing examination of the image as it is processed allows for intuitive decisions. Elements of chance—"happy accidents," as many artists refer to them—can completely alter the image. The results may be interesting, intriguing, sometimes even bizarre. Realization comes when the print says something, perhaps something even better than the initial idea. Deciding to go along with the happy accident is also part of the creative process.

Presenting the image to the viewer is the stage of completion, whether framed, unframed, sewn, or stuffed. This stage is as important as the first because how the image is viewed determines how well it communicates. The viewer has no concept of the artist, the process, or the idea; all the viewer sees is the finished product. The image must communicate the idea on all levels.

The creative process is a never-ending cycle, and one idea builds on another. Images begin to interrelate, and as they mature, further ideas are generated. Keeping one's mind open so that unconscious as well as conscious events are registered will evoke new ideas; for example, pushing a print as far as it will go before it is overworked, scrapping it as hopeless, then returning to it later to salvage some-

thing can contribute to the creative process.

There are no rules for creativity, but there are some guidelines to consider when working with artistic photographic processes.

1. Carefully consider the content of each camera-formed image with which you work. Is the composition strong? Does it evoke some response from you? From viewers? Is it already a good-quality silver print? Do not work with mediocre negatives and expect miracles—they will not save a bad image. Start with the best, and work with the best to get results.

2. Learn to work with large litho film. This is imperative (see Chapter 2). Previsualizing makes the difference between satisfactory results and returning to the darkroom to do the work over.

3. Decide on the size of the finished image. Bigger is not always better. Some images work best large; others need to be small. Do consider large borders for the addition of other photographs, drawings, or collages.

4. Scrutinize each process presented in this book. Study its potential, and evaluate its strengths and weaknesses. Relate each process to the image, asking such questions as: Which surface is best for this particular image? Paper? Fabric? Actual objects such as glass or metal? Should the resulting print be in a single color or in multiple colors? Is it necessary to mix the chemicals from scratch, or is there a commercially available product that will do the same job? What possibilities are open for finishing the image? Will it work to draw, paint, sew, or stuff this par-

ticular image? What are its limitations?

5. Settle in your mind how the finished print will look when it is presented. Shoot for that previsualized image, but be aware that things may happen along the way to influence it.

The list can go on and on. Basic knowledge and some understanding of each process are important bases for making a selection. Explore the entire book first by familiarizing yourself with the illustrations; then read each chapter and examine each process. Underline important information, and write notes to yourself. Question what you read and consider all possibilities. The chances are there to take; the rules are made to be broken.

2 | How to Make the Transparency

USING GRAPHIC ARTS FILM

Just as conventional photography is dependent on continuous-tone films, a major portion of the nonsilver processes use graphic arts films. The main characteristic of these materials is high contrast. They were invented primarily for use in offset printing, lithography, and other forms of mechanical reproduction, and so are referred to as *litho* film, or *high-contrast* film. Graphic arts film can be purchased from Kodak in standard sheet sizes—4 × 5, 5 × 7, 8 × 10, 11 × 14, and 16 × 20—under the name Kodalith Ortho, Type 3, 2556, or 6556. Kodalith 6556 is available in 35mm size (to be exposed as a negative in the camera) in rolls 100 ft. long. It can be ordered specially in wider rolls for darkroom use, or purchased from a graphic arts supply house in offset-printing plate sizes such as 10 × 12 or 14 × 17. GAF, E.I. Dupont de Nemours & Co., Agfa, and others also manufacture this product under their own commercial names.

Sometimes you may find a reduced price on outdated high-contrast film. Buying this is not recommended; it may be fogged because of age and improper storage. Ko-dalith need not be refrigerated, but it should be stored in a cool, dry place. It is rated ASA/ISO 6 to 12 and, because of this low sensitivity to light, it can be worked with under red safelight.

The primary reason to work with this film is to generate a large negative transparency, the same size as you wish the resulting image to be. For example, if you want to make a blueprint 11 × 14 in., the litho film transparency must measure 11 × 14 in. Most of the processes explained in this book use contact printing.

HIGH CONTRAST VERSUS CONTINUOUS TONE

The inherent characteristic of litho film is that it concentrates the shadow and highlight areas into black and white; all middle tones, or grays, drop out. High contrast is created by the type of developer used. Kodalith AB is one high-contrast developer. It can be purchased in a liquid or powder form and prepared as a stock solution. In powder form it makes 2 gallons of stock solution: one gallon of A and one gallon of B. To prepare a working solution, mix one part stock A with one part stock B and *use the combined solution immediately.* If allowed to stand for 2 hours or more, the working solution exhausts itself because of oxidation. An excellent high-contrast developer made by Nacco usually comes in a liquid stock solution of A and B, each to be diluted with water and then mixed 1:1 prior to use.

It is possible to generate a reasonable facsimile of a continuous-tone transparency by using standard litho film and a different developer. Litho film provides denser highlights than conventional film, in addition to higher overall contrast. A standard paper developer (such as Dektol) diluted 1:7 with water brings up the middle grays. Slight overdevelopment darkens the highlight areas further and puts more detail in the shadow areas.

The photographic image you select will dictate which developer to use. A thin, "flat" negative (one with little tonal range) reproduces best on litho film processed in high-contrast AB developer. This generalizes the stronger tonal areas of the negative and cleans up the middle grays. A dense, overexposed negative translates best with litho film developed in Dektol 1:7. This allows for more manipulation, including extraction of contrast in the middle gray areas. The litho film responds to burning in and dodging; so standard photographic printing procedures can produce satisfying results.

For comparison, set up a tray of each developer and make two test strips for each negative: one in AB and one in Dektol 1:7. This comparison allows you to select the right developer for the negative you wish to use. (*Note:* A proportion of 1:7 is *not* the only dilution for Dektol. Some photographers say that a weaker mixture of 1:12 allows longer development times and a better contrast range in the resulting transparency.)

Determine the emulsion side of the litho film by comparing the back to the front. The lighter side is the emulsion side.

GENERATING THE TRANSPARENCY

Any conventional darkroom used for black-and-white printing will do to make the large transparency. Prepare the developers for the desired results. Set up an acetic acid stop bath and a tray of standard fixer with a hardener. Use a *red* safelight.

Turn off the room light and place the continuous-tone original, in this case a negative, in the enlarger. (A positive color slide can also be used to save the later necessary contact printing stage, as will be described.) Be sure the original is free of dust and fingerprints. The red safelight should be at least 4 ft. (1.2 m) away from the developing tray or the negative may be fogged. Set the easel for the film size (4 × 5, 8 × 10, etc.) and focus with the lens wide open. When ready, stop the lens down to its smallest aperture—ideally f/16 or smaller. The smaller the aperture, the greater the definition; a longer exposure time also allows all the tones to transmit in the correct relationship to the continuous-tone original. At the other end of the scale, a wider aperture produces a higher contrast in the image, with less definition and a narrower tonal range.

Cut a sheet of film into large strips and place one in the easel, emulsion side up. The emulsion side of the film is usually lighter in color than the acetate back. Kodalith's emulsion side is light pink, and the backing dark purple. Make a test strip using 3-second intervals. This is only a suggested place to start; the actual time depends on the density of the projected negative.

Use a test strip to determine the correct exposure time.

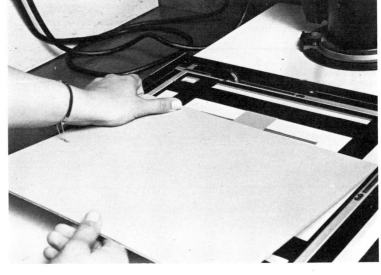

Expose each section, building up a gradation of exposures.

PROCESSING THE TRANSPARENCY

For high contrast, develop the test strip in AB for a minimum of 2 minutes with constant agitation; for middle tones, use paper developer (Dektol 1:7) for a minute to a minute and a half. When development is complete, immerse the strip in a stop bath for 30 seconds. Fix until the unexposed emulsion (or light pink color) has disappeared. The highlights will be completely clear.

(**Note:** Always handle the film carefully in the processing stage because the emulsion is quite soft and vulnerable to scratches and fingerprints. Tongs tend to scratch; so it is best to use rubber gloves if photographic chemicals irritate your hands. *Be sure the gloves are completely dry before you handle unexposed film.*)

Evaluate the image under *room light.*

For high contrast, choose those exposure and developing times that render the shadow areas completely opaque. In other words, you want the best separation of shadows and highlights into distinct black areas and clear areas on the film.

For continuous tones, select exposure and developing times for a result that is similar to a normal black-and-white print. There should be some detail in the shadows, and tonal ranges should correspond closely to those of the negative projected. Burning in and dodging may be used if areas need more or less light to render a satisfactory image; litho film is sensitive, to a degree, to these procedures.

After evaluating the test strip, expose a full sheet of film at the time selected.

Immerse the entire exposed sheet emulsion-side up in the developer. Agitate it continuously for the full recommended development time.

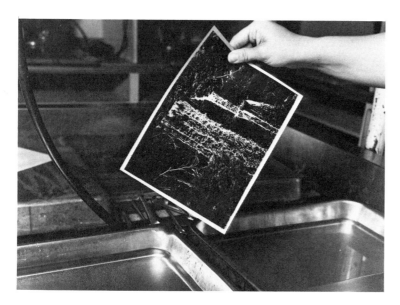

Remove the sheet from the developer and allow the excess developer to drain back into the tray.

Immerse the film in stop bath for 30 seconds, and then drain off the excess.

Place the sheet in the fixer and agitate continuously.

Remove the film when the backing has cleared and evaluate the exposure in room light.

After the film has been fixed, wash it in running water for 10 to 20 minutes. The use of a fixer neutralizer such as Hypo Clear will cut the washing time in half and save water, as well. A tray or similar equipment is best; do not use a print washer or any washing machinery that might scratch or damage the film. Prepare a tray of Photo-Flo or other wetting agent according to the package directions. Immerse the film. Hold the film by one corner to drain off the excess, and hang it to dry. *Do not squeegee the film—this causes scratching.*

At this point, if the projected original was a negative, the resulting litho image is an enlarged positive. To get a negative, contact-print the positive with another piece of film. But first retouch the positive.

RETOUCHING THE TRANSPARENCY

After the transparency is completely dry, check it for dust spots, "pinholes" (small, clear dots usually caused by uneven processing), and scratches. On high-contrast transparencies, these flaws can be easily repaired by a water-soluble paint-on solution such as Kodak's Opaque, which is available in two colors—red and black. The red is preferable because it can be seen more easily. Use a no. *00* or no. *000* brush to eliminate flaws *on the emulsion side.* Opaque dries quickly, and because it is water-soluble, any errors in application can be wiped off with a soft cloth or washed off under running water. If water comes in contact with a surface that has been opaqued in any of the following processes, the transparency can be ruined. Keep a retouched transparency in a dry place.

Opaque can also be used to fill in or alter whole areas of a transparency, for example, to eliminate undesirable backgrounds or subject matter (see illustration).

Exposed areas can also be removed by etching with a small, sharp artist's knife. This is a tricky maneuver because there is a fine line between removing a spot and damaging the acetate film base. Work carefully on a light table, etching in quick, even strokes on the emulsion side of the film. Practice makes perfect, so begin with a piece of scrap film.

For a continuous-tone transparency, india ink or Spotone retouching solution works well. Using a fine paintbrush, stipple pinholes and fill in dust spots. Slightly diluting the solutions and using a larger brush allow for filling in larger areas such as weak shadows. Retouch any areas lacking detail. All this repair work must be done prior to contact printing, and all surfaces must be completely dry before the next step.

Using a No. 000 brush, paint out or retouch problem areas with opaque.

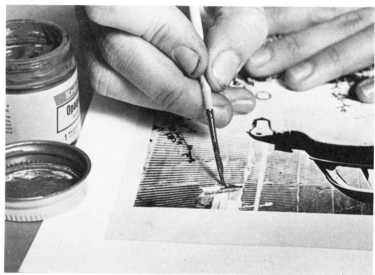

When using a retouching knife, etch in even strokes and be careful not to damage the acetate support.

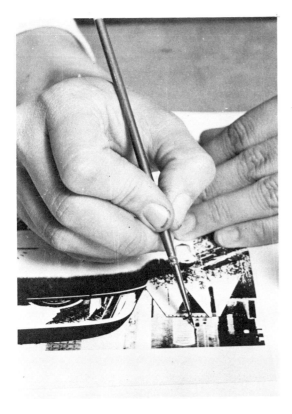

Stiple the film using a No. 000 brush to create small dots of Spotone.

CONTACT PRINTING

The standard method for contact printing usually eliminates the need for making a test strip. Simply set the enlarger and the timer at the same position and settings required to create the positive. Place a fresh sheet of film, emulsion side up, on the baseboard of the enlarger. Set the positive on top, and cover both sheets with a piece of heavy, clean glass to ensure complete contact. (A contact-printing frame can be used for precise registry.) Expose the film, using the same time you used to make the negative. Process in the same developer you used for the positive—either AB or Dektol 1:7—using the same development times. Stop, fix, and evaluate.

TEST-STRIP TECHNIQUE

If the resulting enlarged negative seems too thin or too dense, you may wish to go back to the test-strip method to determine the correct contact-printing exposure. Set the aperture of the lens at *f*/16 or smaller. Set the timer at intervals of 3 seconds, for a start. Place a fresh sheet of film (or. cut a sheet into smaller strips), emulsion side up, on the baseboard or in the print frame. Lay the positive on top and cover it with glass. Using a piece of heavy cardboard or opaque paper, cover all but a 1-in. (25-mm) strip of the left side of the image. Expose this visible segment for 3 seconds, then move the cardboard over an inch and expose a new area for 3 seconds. You are now building sections of the exposure in 3-second intervals, one of which will be the proper exposure for the entire image.

Process the completed test strip in the same developer you used for the positive. Stop, fix, evaluate, and select the correct exposure time. Expose a full sheet of fresh film at this interval. Burning in and dodging can also be done in the contact-printing stage, to control specified areas selectively. Process the film. Immerse in Photo-Flo and hang it to dry. You now have same-size negative and positive transparencies with which to explore various nonsilver processes.

HALFTONE TRANSPARENCIES

There is a third choice available when working with line film which creates a completely different image. *Halftone*—the term used by printers and graphic artists—refers to the conversion of a continuous-tone image to a pattern of small dots. Shadow areas are composed of many dots clustered together, and highlight areas are made of fewer well-spaced dots. With a magnifying glass examine a photograph from a printed source such as a magazine or newspaper and you will see the halftone or "screened" image as it has been converted into a pattern of dots.

This dot pattern is important in offset or mechanical printing. If continuous tones were actually used, the image would fill in with ink. Each halftone dot, however, prints separately; but the sum of the dots suggests a continuous-tone image. Halftone dots can also make interesting patterns that can be incorporated into the final concept of an image.

There are two ways to produce a halftone image. One way is through the use of a halftone screen with high-contrast litho film and AB developer. The second, and by far the simpler, method is to use an already commercially screened emulsion, such as Kodak's Auto-screen. This film, as well as half-

Use cotton handling gloves when working with halftone contact screens.

tone screens, can be purchased at graphic arts supply stores. Autoscreen is available in sizes from 4 × 5 up to 11 × 14 and has a 133-line-per-inch dot pattern. Separate halftone screens come in various sizes and line ratings, which refer to the number of dots per inch per line. For example, a 65-line screen has a dot pattern of 65 lines per inch and is considered a coarse screen for halftone work. A 133-line-per-inch dot pattern is an average screen for work with these processes. Selection of the proper screen is essential; the result you wish to achieve determines the size and type required.

HOW TO MAKE A HALFTONE TRANSPARENCY USING LITHO FILM

This is a simple procedure that requires an enlarger, litho film, a halftone screen, and AB developer. Place the continuous-tone negative in the enlarger. Adjust for the desired size, focus at full aperture on the baseboard, and stop down to *f*/16. The size of the aperture affects the overall image; a small

aperture accentuates the dot pattern and allows for more separation of tones, whereas a larger aperture produces an image of higher contrast. The AB developer should be freshly mixed 1:1, and all other chemicals should be set up and ready to be used.

For projection work with a continuous-tone black-and-white negative, Kodak recommends a Magenta Contact Screen (positive). The end result from this first step will be a positive screened image. The 133-line-per-inch screen is the average dot size to start with. Kodak publishes two excellent pamphlets, *Halftone Methods for the Graphic Arts*, Q-3, and *Contact Screens: Types and Applications*, Q-21, which explain in depth the information given briefly here. Also see Appendix A, which tells how to make a halftone screen.

Place a piece of litho film, a test strip or a full sheet, emulsion side up on the baseboard. Using cotton handling gloves to handle the screen, lay the Magenta Contact Screen on top, emulsion side down. A vacuum-printing frame is best, but a piece of clean, *heavy* glass will do (see illustration). Set the glass over the sandwiched film screen. Set the timer for five sec-

onds. This is only a starting time; the process may take longer since you are exposing the image through a screen. Make a test exposure at 5-second intervals. Remove the glass and carefully put the contact screen away, because it can be easily scratched. Process the line film in AB developer for a minimum of 2 minutes with constant agitation. Stop and fix.

To evaluate the resulting halftone positive test strip, find the sections in which all the individual dots composing the image are completely opaque. If the shadow areas are completely filled, the film has been overexposed; if the highlights are completely clear or the dots are small and uneven, it is underexposed. Select a time that produces an overall even pattern of dots, visible in both the highlights and the shadow areas.

To achieve a wider tonal range, add exposure to the shadow areas to ensure a pinpoint dot pattern. This is accomplished by "flashing" the sandwiched litho film and contact screen. Remove the negative from the enlarger after making the test strip, and expose the latent image with a flash of light. Diffuse the enlarger light with a handkerchief or white paper over the lens, which

Place the film emulsion-side up on the baseboard, cover it with the contact screen and use a piece of heavy glass to secure it if a vacuum frame is not available.

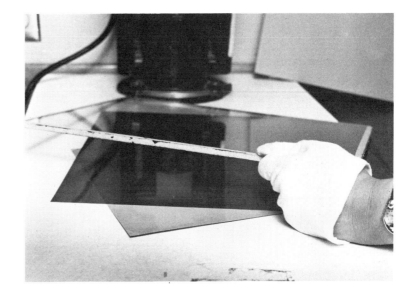

An example of a good highlight dot.

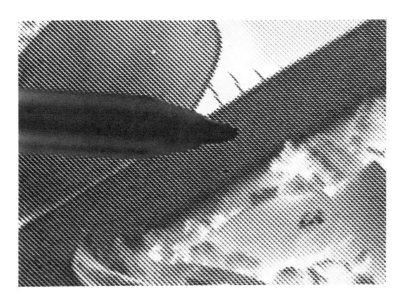

Cover enlarging lens with a handkerchief or thin white paper and secure it with a rubber band.

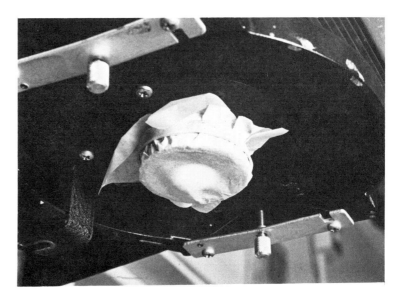

A flash test strip saves time by revealing a wide range of possible exposures. The grid pattern apparent in the print shown here resulted from two series of exposures made perpendicular to each other. The first was completed while the negative was in the enlarger. For the second exposure the negative was removed, the enlarger light diffused, and the test strip "flashed." The print therefore allowed the photographer to evaluate a variety of different combinations of flash and direct exposure.

allows for a more even flash, and stop the lens down completely (see illustration on previous page). The flash exposure affects only the shadow end of the scale; the highlight areas remain unchanged. Flash the test strip for 5-second intervals moving at right angles to the direction in which you made the initial test strip exposure. Make a flash test strip to determine the correct time and achieve the desired dot size.

After evaluating the test strip, make the halftone positive. Process and wash it. Use a wetting agent, and allow the positive to air-dry. When the halftone positive is completely dry, do any necessary retouching, and contact-print it with another piece of litho film to make a negative. *Do not* reuse the halftone screen in this step, or the dots will create a moiré (a wavy or circular pattern created by the interaction of light rays refracted through the halftone screen and the halftone positive). You now have both a halftone positive and a negative transparency with which to work in the nonsilver processes described later.

HOW TO MAKE A HALFTONE TRANSPARENCY USING AUTOSCREEN

This is by far the easier way of making a halftone transparency, because Kodalith Autoscreen Ortho film has a 133-line-per-inch dot pattern built into the emulsion layer. The built-in dot pattern increases clarity and sharpness because you do not expose through a contact screen. Autoscreen, however, is expensive. If you plan to make many halftones, it might be more economical to purchase a halftone screen.

Use Kodalith AB developer to process Autoscreen. It is advisable to purchase film the same size as you wish the finished image to be.

The largest Autoscreen size is 11 × 14. This limitation can be overcome by making a small positive image on Autoscreen 4 × 5 film and projecting the resulting positive on a larger sheet of standard litho film. Keep in mind, however, that the dots when projected will be larger, and a coarser overall dot image will result in the negative. This can create an effect you might not want.

Place the continuous-tone negative in the enlarger. Set the size, focus at full aperture, and stop down to f/16. Make an Autoscreen test strip at 5-second intervals. Process in AB with constant agitation for a minimum of 2 minutes. Stop and fix. Evaluate the test strip the same way you would if you had used a halftone screen. If flashing is required, it can be done in the same manner as with a contact screen. Select the correct time. Expose a fresh sheet of Autoscreen. Process and wash. Use a wetting agent and hang to dry.

To achieve a halftone negative, contact-print the positive with standard litho film. Do not contact-print the Autoscreen with another piece of Autoscreen or the dot pattern will moiré.

OTHER ADVANCED TECHNIQUES

Litho film has many of the same characteristics as black-and-white photo paper. It is sensitive to the Sabattier effect—or, as many incorrectly refer to it, solarization. It can also be toned, bleached, and otherwise manipulated.

Darkroom Art by Jerry Burchfield (Amphoto, 1981) refers to many of these litho film techniques. Burchfield presents precise information on how to generate tonal separations, step-by-step instructions for posterization, and an excellent "make your own" registration method for small-format work with litho film. All these techniques can be applied to a larger litho film format for application in nonsilver work.

Color separation of a transparency or reflective copy is a detailed and precise process requiring special equipment and some advanced experience in graphic arts. Arnold Gassan's *Handbook for Contemporary Photography* (Light Impressions, 1977) best describes how to color separate from 35mm transparencies requiring the

use of *only* a darkroom, Kodak Wratten gelatin filters, and Kodak Super-XX film.

Kodak's *Basic Color for the Graphic Arts*, Q-7, presents a simple step-by-step explanation of color principles, color reproduction, and color separation from both transparent and reflective copy. These instructions require the use of a process camera, darkroom, halftone screens, and other graphic arts equipment.

To understand color separation fully requires extensive reading and actual hands-on practice before satisfying results can be achieved. The alternative is to pay for the service from a commercial source or printing firm that specializes in high-quality color separations.

STEP-SAVING OPTIONS

The methods described thus far are the standard procedures most commonly used. There are step-saving alternatives worth knowing about, however. Using Kodak High Speed Duplicating Film 2575 is an option that eliminates the final contact-printing stage when either a high-contrast or a continuous-tone image is the desired result. This black-and-white film is expensive, but it is a time saver. It is available in standard sheet film sizes, and when exposed it translates a negative into a negative or a positive into a positive. Therefore, a high-contrast negative from a continuous-tone negative can be produced in a single step.

High Speed Duplicating Film 2575 responds to light in the opposite way from regular litho film because it is a reversal material. High Speed Duplicating Film has been

preexposed, so the more light the film receives the less density is recorded and vice versa.

(**Note:** Dodging makes selected areas *darker*, and burning makes areas of the image *lighter*. Handling the film is basically the same as for Kodalith Ortho Type 3, and the choice of developers depends on the desired result. Use AB for high-contrast and Dektol 1:7 for continuous-tone prints.)

Another alternate method is to contact-print 35mm negatives directly onto litho film, making a transparent positive proof sheet. Develop it in Dektol 1:7. When the sheet is processed and dried, select a positive image and enlarge it onto litho film to create a larger negative. There may be some loss of detail in this process, and any flaws that appear during contact printing, such as dust, are magnified.

Still another option is to expose, in the camera, a continuous-tone fine-grain film (sizes 135, 120, or 220 only), Panatomic X (ASA/ISO 64 for tungsten lights and ASA/ISO 80 for daylight). Purchase a Kodak Direct Positive Film Developing Kit and follow the directions for reversal development of Panatomic X. The result is a black-and-white continuous-tone *positive*, which can be enlarged directly onto litho film to create a large negative.

Building a library of enlarged negatives and positives suitable for nonsilver processes is similar to increasing your vocabulary. Images soon begin to relate and interchange, and you will begin to develop a style all your own.

3
ALTERNATIVES TO MAKING YOUR OWN TRANSPARENCIES

Making your own pinhole camera is simple. You will need to experiment to find the proper exposure, but with enough experience you can generate fascinating images. The photograph shown at left was created by Laura Aguilar and is entitled Cool Raul.

If lack of access to a darkroom prevents you from making your own transparencies, there are alternatives. The most obvious one, of course, is to pay someone else to do the work in a commercial lab. A second option is the use of already made or "found" negatives. Third, you can directly expose litho film in a homemade pinhole camera to create large negatives with very different effects. A fourth choice requires incorporating the information presented in Chapter 10 on magazine lifts with the contact-printing process of litho film. Any of these alternatives can be used by itself or combined with other methods to make transparencies.

THE PROCESS CAMERA

A commercial graphic arts processing lab, or "stat house," as it is often referred to in the trade, is a photographic lab equipped with a large "copy" or process camera that can quickly and easily do many of the procedures discussed in Chapter 2. The process camera is a large camera that usually occupies two rooms. One room houses the lens, lights and copy board, and the other is a small darkroom where the technician operates the controls, handles the film, and does the processing (see illustrations on following page).

The original is referred to as "copy." It should be clean, and it should include strong contrasts, detail in shadows, and the overall quality of a first-rate black-and-white print. A high-contrast negative is referred to as a "line shot" or "line copy." A halftone negative is "screened" to acquire its characteristic dot pattern. (*Note:* You will need to know the desired dot size; 133 lines per inch is preferable as was discussed in Chapter 2.)

Standard copy cameras usually have the capability of enlarging and reducing originals from wallet size to 20 × 24. The camera accepts film sizes as small as 4 × 5 and as large as 20 × 24. The enlargement or reduction size of the negative you get is computed in percentages. Shooting copy same size is 100 percent. Reducing copy to half the original size is 50 percent, enlarging it to twice its size is 200 percent, and so on. The computation is done by measuring one side,

27

Flat reflective copy is placed on the copyboard behind glass, and then swung forward, perpendicular to the lens. Lights are at a 45 degree angle to copying surface.

The enlarged or reduced image is projected onto frosted ground glass, and measuring insures the correct percentage of the reduction or enlargement.

Processing of litho film is completed on the other side of the camera in a fully equipped darkroom.

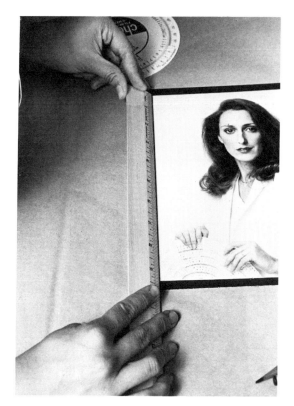

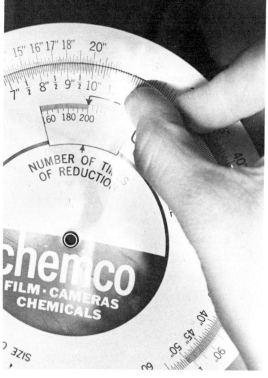

Be sure to measure the original accurately so the computation of percentage is correct.

Place the original's measurement opposite the desired size, and the dial will indicate the percentage of reduction or enlargement.

such as the height of the original, and dialing in that measurement opposite the desired height of the resultant transparency. This lining up of original size against desired size gives the correct percentage of enlargement or reduction for the entire image. It is always a good idea to compute the new size for both the *height* and the *width* of each original to compare and determine if the transparency will be in correct proportion.

If you need a lot of work reduced or enlarged, it is best to purchase a proportion wheel for your own use. Using my own wheel, I make a layout on a large sheet of bond paper, reducing or enlarging the originals to the correct sizes, to obtain a working idea of how the new image will look after all camera work is complete. Showing this

Mark the side measured, its size in inches, and the percentage of enlargement or reduction.

sketched layout to the technician aids me in getting what I want done.

Reducing and enlarging usually cost an extra setup charge, so supplying originals that can be shot to the same size (100 percent) saves money. Prices are also computed in relationship to standard film sizes. Therefore, working with the formats of 4 × 5, 5 × 7, 8 × 10, 10 × 12, 11 × 14, 12 × 18, 14 × 17, 16 × 20, 18 × 23, or 20 × 24 is advisable. Determining reductions or enlargements in relation to film sizes also saves money.

Give the technician all the information you believe is important. Be sure to describe how you plan to use the transparency, such as whether you plan to make a blueprint or brownprint. Do not leave the shop until you believe the technician understands what you need. Before the job is begun, ask if you may have, or purchase, any mistakes—that is, shots ruined by uneven development, partially fogged film, irregular dot patterns, and so on—that occur on your job. Sometimes the accidents are very interesting to you, though to the camera technician they are just material to be discarded.

FOUND NEGATIVES

There are unlimited sources for negatives if you choose not to make your own. One source I term "found negatives," and the most obvious place to look is "right in your own backyard." Check drawers, trunks, storage boxes, and family photo albums for old continuous-tone negatives. Elderly relatives seem traditionally to save everything, so check for large-format negatives such as 2¼ × 2¼,

2¼ × 3¼, and 4 × 5. Regular 35mm black-and-white negatives and color negatives from the 1940s or later, if they are well preserved, will do, but some experimentation is required to obtain satisfactory results with them. Another place to look is antique shops, which now deal heavily in photographic paraphernalia. Glass negatives may be easily acquired, as well as the larger-format sizes of film. Garage sales and swap meets are also excellent spots for good buys and unusual finds.

There is a great deal of adventure in working this way. It is like going through someone's closet or sneaking a peak at a passage in a diary. Moreover, it is a serendipitous way to acquire images. Although it requires a great deal of thought to use photographs of the past, the blueprint and brownprint processes especially lend themselves to working with vintage images.

Do not overlook vintage prints. These positive images can be copied by a process camera or contact-printed directly onto litho film to create negatives. Contact printing requires a darkroom, trays, chemicals, and a reliable light source, preferably an enlarger. See Chapter 2 for the step-by-step procedure for contact printing. (*Note:* Contact printing through the fiber base of the paper generally requires relatively long exposure times; opening the lens by as many as three stops from f/16 to f/5.6 expedites this part of the process.

A third category of found negatives is medical and dental X rays. Junk shops, swap meets, and even garage sales sometimes have them. Old medical buildings due for demolition may have auctions where large boxes of X rays can be

found mixed in with office furniture and file cabinets. Asking your own physician or dentist for your X rays—if they are no longer needed—is your best source. (*Note:* Sometimes lab technicians reshoot X rays and discard the old ones—unless you ask for them.)

In working with any found negatives, unless the subjects are your own family, it is best to "sufficiently alter" or rework them—not only to personalize them, but also to avoid potential legal problems. As a protective measure against lawsuits that could arise, consult an attorney before you display any images generated by found negatives.

PINHOLE CAMERA

Without the use of a conventional darkroom, with just a cardboard box, a dark closet, and some developing trays, you can generate imagery for nonsilver processes by shooting line film in a pinhole camera. The principles of the pinhole camera, which had been known long before Daguerre made the first photograph, can be utilized with line film to create interesting large-format negatives. The primary advantage of the pinhole camera is the immediacy it provides. Once all the kinks are worked out, you can conceptualize an idea, shoot large-size litho film right in the pinhole camera, process it, and evaluate it.

The most concise information to date on pinhole photography can be found in the Kodak *Encyclopedia of Practical Photography*, volume 11. The section titled "Pinhole Camera" discusses pinhole principles, camera construction, and general exposure times for different ASA films. You must be familiar with this information if you wish to be precise in using the pinhole camera as a means for making large negatives. The following general information is but a starting point. Trial and error are required to master the art of pinhole photography.

CONSTRUCTING THE CAMERA

You need an ordinary cardboard box, such as a round oatmeal container, shoe box, or large packing box. Paint the interior of the box with flat black paint. This reduces reflections and the bouncing of light within the camera body. Make sure that the back, where the camera will be loaded with film, can be sealed light-tight with black tape. Black photographic tape, electrician's tape, or Mystik tape will work.

Next, cut a small opening about 2 in. (50mm) square in the center of the front panel. This is where the pinhole will be placed.

THE PINHOLE

The simplest procedure for making the pinhole is to pierce, with a sewing needle, a piece of heavy-duty aluminum foil or a metal pie tin. A no. 10 needle creates a pinhole with a 75 to 150mm focal length; this means the film should be 3 to 6 in. (75 to 150mm) away from the pinhole for maximum detail and exposure control.

Make sure the pinhole is a clean puncture. Pierce the foil or pie tin, rotating the needle as you push, and stop when half the needle's shank is through the metal surface. Remove the needle and check the hole with a magnifying glass for any ragged edges.

Center the pinhole over the 2-in. (51-mm) front opening of the camera body, and tape it in place with black tape. *Make sure there are no light leaks.*

Make an opaque black flap to cover the pinhole opening. This is the shutter, which is removed for making exposures.

EXPOSING LITHO FILM IN A PINHOLE CAMERA

Exposure times vary, depending on camera construction, lighting situations, and time of day. A series of tests is advisable. For example, a time to start with, in bright sun using Kodalith Ortho Type 3 in a 3- to 6-in. (75- to 150mm) camera with a pinhole formed by a no. 10 needle, is approximately 90 seconds. (Refer to the Kodak *Encyclopedia*, volume 11, for how to compute exposure times.)

To load the camera, open the back under red safelight. Place the litho film, cut to fit if necessary, emulsion side toward the pinhole, in the back of the camera lid, and secure the film's corners with masking tape. Place the lid back on the camera body and seal it shut with black tape so that it is light-tight.

Make sure the shutter flap is securely in place over the pinhole. Proceed outdoors with the camera and place it on a steady surface, such as a brick fence, hood of a car, or table. A piece of double-sided tape can be helpful in anchoring it. Because exposures usually take several minutes, the camera must remain stationary.

Remove the pinhole cover, and using a watch or portable kitchen timer, expose the film for the estimated time. When exposure is complete, cover the pinhole open-

ing and return with the camera to the red safelight area for unloading and processing.

Remove the film from the camera back and process the litho film as described in Chapter 2. Remember that Kodalith Developer AB renders a high-contrast image and Dektol 1:7 (or 1:12) produces a continuous-tone transparency. Evaluate the results. Expose more film if necessary, compensating for changing light situations and under- or overexposures.

Ordinary photographic paper can be substituted for litho film to make a paper negative. Use single-weight fiber (not RC) and normal grade no. 2 paper, and develop it in Dektol 1:2 for 2 minutes. Fix and evaluate.

Once you determine the correct times for either film or paper and the various lighting conditions, you can pursue some of the creative advantages of this method.

PINHOLE CAMERA VARIATIONS

Because exposures are calibrated in minutes, ghostlike images of people can be created. For example, have a person hold still for half the exposure time, then have him or her move slowly across the picture area. The motion causes an eerie blur.

Placing the film in the camera slightly bowed or otherwise bent distorts the resulting negative. Distortions are especially useful in fabric printing because they can be further exaggerated by sewing and stuffing.

Multiple pinhole cameras can be constructed by piercing more than one hole in round hatboxes, 5-gallon ice-cream containers, or large sectioned boxes. Sensitizing three-dimensional objects, like ceramic bowls, with Rockland's Liquid Light (see Chapter 9) and placing them in a multi-pinhole hatbox camera would create on the surface of the bowl a continuous-tone panorama of whatever scene was being photographed. You can also use litho film in a pinhole hatbox camera; form it in a big loop, secure it into a circle with tape, and place it inside the camera to expose a 360-degree panorama. The variables are unlimited, and the ideas are endless in pinhole photography. *Experimentation is important!*

MAGAZINE LIFTS

A third method of acquiring negatives requires the incorporation of a technique—magazine lifts—and the contact-printing process of litho film. Magazine lifts, the "lifting" of the printed inks of magazine images into a flexible plastic support, such as clear Con-Tact paper or gloss polymer medium are discussed in detail in Chapter 10.

Magazine lifts are semi-transparent positive images that can be either directly exposed with nonsilver sensitized materials to create negative photographs, or

contact-printed onto litho film to make negatives. Either method renders consistent results, but the latter allows the opportunity to alter the density of the resulting negative. Over- or underexposing, burning, and dodging during the contact-printing stage improve the overall quality of the magazine-lift image. Sometimes such a generation loss cleans up highlights, and, of course, the selection of litho film developers determines whether the resulting negative is high contrast (AB) or continuous tone (Dektol 1:7).

Magazine lifts made with gloss

Multiple pinholes create multiple images. The photograph below was made by Laura Aguilar in a pinhole camera constructed from a hatbox. The artist curved litho film she used to make the original negative to fit the hatbox.

polymer medium can be stretched and distorted. The degrees of distortion can be recorded by contact-printing each variation with litho film. Sequential film negatives of such a progression create just another of many applications of this transparency alternative.

Contact-printing litho film with newspaper and magazine tear sheets; single-weight prints, and paper negatives are all viable choices for generating large negatives. Your own black-and-white prints on single-weight paper can easily be transposed onto litho film. Paper negatives made in a pinhole camera can also be contact-printed both on litho film or directly on nonsilver emulsions.

PHOTOGRAMS

Three-dimensional objects placed directly on sensitized surfaces produce photograms of excellent quality. Creating photograms on litho film allows you to duplicate an arrangement over and over again in a multiple edition. Retouching the film transparency allows manipulation of the spatial relationships created by the photogram process. Drawing on the film with india ink or opaque ink can complete contours and further add to spatial illusions; airbrushing, scratching, and other forms of altering can also enhance the photogram. (*Hint*: Wetting the objects with developer and then placing them on the film before exposing it to light usually renders the texture of the surface touching the film.)

Do not forget that *you* are three-dimensional! Lying on a sensitized surface or on litho film is a way of creating an interesting self-portrait.

The incorporation of your silhouette, hands, or face along with special objects further personalizes such a image.

Finally, clear or frosted sheets of acetate can be painted or sketched on with india ink. They can also be marked with a grease pencil. Lettering, pen and crosshatch drawings, graphics, and strongly painted tonal washes reproduce faithfully when contact-printed with these artistic processes. In fact, two of these light-sensitive printing methods—blueprinting and brownprinting—were used primarily for the reproduction of graphs and "blueprints" around the turn of the century; hence the terms *blueline* and *brownline*.

You need not feel limited by a lack of darkroom facilities, as you will understand after reading the following chapters.

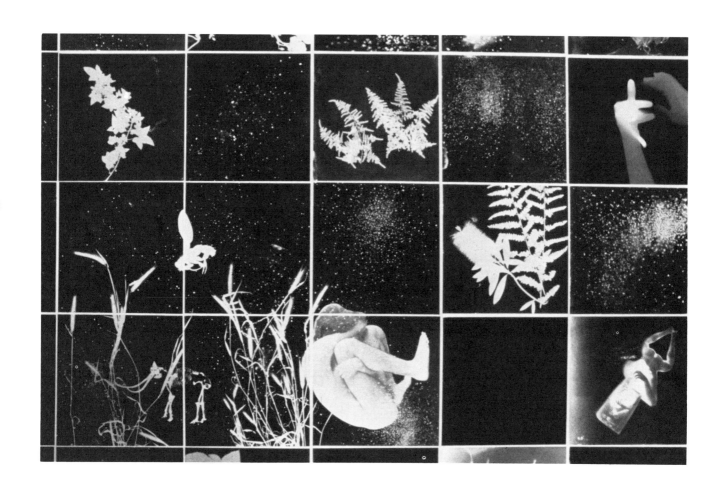

Shown here is a detail from Sheila Pinkel's Cyanotype Mural. *The 8- x 20-ft. mural was commissioned for the foyer of the Park La Brea Towers in Los Angeles. Pinkel combined photographic negatives with photograms to create the gridded image.*

4

CYANOTYPE

Blueprinting is the process most people try when beginning nonsilver photography. It is the introductory course to paint-on emulsions, a sort of "Nonsilver 101." Moreover, measuring dry chemicals, coating paper, making paddle brushes, exposing paper or fabric to ultraviolet light sources, and dipping fabric are skills that should be mastered for blueprinting as well as for the other processes discussed in later chapters.

Blueprinting or "cyanotype" is the oldest nonsilver photographic printing process. This ferric salt emulsion, invented by Sir John Herschel (1792–1871), was used by such early photographers as Gustave Le Gray (1820–1862) and Clarence White (1871–1925). Blueprinting saw a revival among contemporary photographers in the early 1960s. Robert Fitcher, Bea Nettles, and Darryl Curran have incorporated it in their work and are responsible for its introduction into the curriculum of many university and college photography and art departments.

Blueprinting yields a cyan image on a white background. Matching the content of the image with this process requires a subjective decision, but the color value of the blue itself seems to dictate the selection of cool, serene imagery.

To blueprint an image, you need a negative or positive transparency the same size you wish the resulting image to be. Chapters 2 and 3 present various ways of making such transparencies.

BLUEPRINTING FORMULAS

There are two standard formulas for blueprinting. Formula I renders an intense, brilliant cyan; it should be prepared just prior to its use and discarded after the printing session. Formula II, on the other hand, renders a less intense cyan color, but if stored in separate brown bottles it can be used for several months.

FORMULA I

1.25 g oxalic acid
33.7 g ferric ammonium citrate
(green scales or powder)
11.2 g potassium ferricyanide
.25 g ammonium dichromate
250 cc distilled water

Prepare just before use, and discard after the printing session.

FORMULA II

Solution A
50 g ferric ammonium citrate
(green scales or powder)
250 cc distilled water

Solution B
35 g potassium ferricyanide
250 cc distilled water

Mix A and B just before use

Caution: All these chemicals are poisonous in both crystalline and liquid states. They are toxic if swallowed and also irritate skin. WEAR RUBBER GLOVES WHEN YOU MIX AND HANDLE THESE CHEMICALS.

In each formula the ferric ammonium citrate is responsible for the cyan color and the potassium ferricyanide is the primary sensitizer which becomes active when exposed to ultraviolet light. Formula I uses ammonium dichromate, which is also sensitive to ultraviolet light. This additional chemical is responsible for the more brilliant cyan color, but it is also the reason this formula is exhausted within 24 hours of preparation.

MEASURING AND MIXING THE CHEMICALS

Since accuracy insures proper results, use precise measuring devices. You need a gram scale to weigh all the chemicals used in these and other processes discussed in this book. Be sure the scale is durable as well as sensitive to minute changes in weight. Costs vary from brand to brand. Gram scales can be purchased at chemical supply stores, teacher's supply houses, and some photography supply stores. (See Appendix D, "Sources of Supply.") Understanding how this device works is important to ensure accurate measuring. Therefore, take the time to read the directions carefully and familiarize yourself with the scale's operation. Place a lightweight piece of card stock or cardboard on the measuring tray (see illustrations). Place weights totaling the weight you need opposite the measuring tray, then slowly measure out the dry chemical until the scale is balanced. Use a fresh piece of paper for each chemical, and always return the scale to zero before you measure the next amount.

Balance the scale.

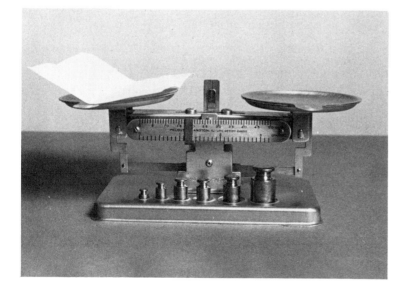

Place the weights in the desired combination on the right side of the scale. The example here shows 35 g.

Carefully measure out the first chemical onto the folded card.

PREPARING FORMULA II

This mixture is preferrable because of its relatively long shelf life, and because larger quantities of A and B can be made, stored, and readied for use very easily.

Do all measuring and mixing of the chemicals used in Formula II in subdued room light, or in yellow safelight conditions; these chemicals are sensitive to normal room light in both the crystalline and liquid states.

You will need two pieces of lightweight card stock or cardboard, one for each chemical.

Measure 50 g of ferric ammonium citrate, and set it aside for mixing later. Now place the other piece of cardboard on the scale and measure 35 g of potassium ferricyanide. Put it aside also.

Pour into a large glass or plastic mixing container (a kitchen measuring cup will do) 250 cc of distilled water. You must use distilled water because the ferric ammonium citrate will react to the metallic content of ordinary tap water. Slowly add 50 g of ferric ammonium citrate to the distilled water as you stir the solution vigorously.

After all the dry chemicals have been added, stir for about 3 minutes, or until the solution is thoroughly mixed. Pour this liquid into a brown bottle marked "Blueprint—Solution A."

Clean all utensils. Measure 250 cc of distilled water, and repeat the same mixing procedure for the potassium ferricyanide. Make sure it is thoroughly mixed and pour it into a brown bottle labeled "Blueprint—Solution B." Date each bottle so that you will know its shelf life and can anticipate when the solutions will be exhausted.

Cyanotype

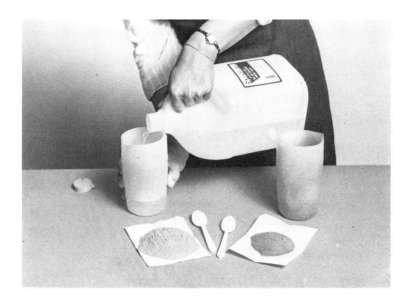

Use distilled water. Pour equal amounts of water into two containers.

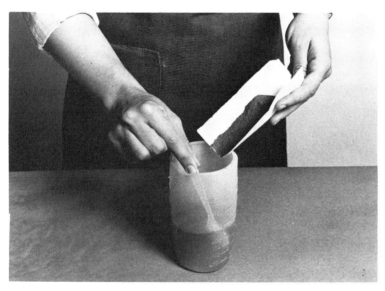

Slowly add the ferric ammonium citrate crystals to the water, stirring constantly.

Pour the completely dissolved solution into bottle marked Solution A.

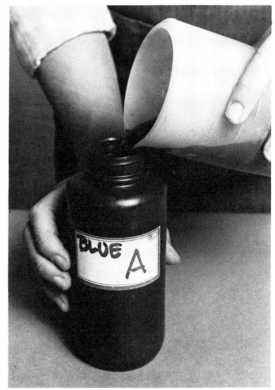

Repeat the procedure for potassium ferricyanide, and pour this second solution into its own marked bottle.

MIXING FORMULA I

Measure the four chemicals separately. Divide the distilled water into four equal parts in four separate containers. Mix each chemical in its own container of water, then combine all four in a brown bottle. Remember: This solution *must be discarded after 24 hours.*

PREPARING TO BLUEPRINT

Applying blueprint solution to paper is quite easy if you follow these guidelines.

PAPER SELECTION

The paper selected should be able to withstand heavy applications of liquids. It should also withstand long periods of immersion in running water. The best papers for this purpose are the ones used for watercolor painting, such as Rives BFK or Arches, which contain a high percentage of rag fiber and are known for their durability. Paper sizing is not necessary, but should yellow staining occur, you should read the section on sizing in Chapter 6.

APPLICATION TOOL

You need a good-quality sable or camel's hair brush, found in art supply stores, to apply the blueprint emulsion. The brush size depends on the size of the area you wish to cover. Instead of a brush, however, try an inexpensive sponge applicator brush, available at hardware stores (see illustrations on following page). Tear off the sponge, and you have a flexible plastic paddle. Wrap a small piece of felt or other soft fabric around the end of this paddle, and secure it with a rubber band. This home-made tool, also referred to as a Blanchard brush, evenly coats the surface of the paper without visible brush strokes like those from a paintbrush. Now, with masking tape, secure the paper to a piece of glass or another clean, hard, non-absorbent surface on a sturdy tabletop.

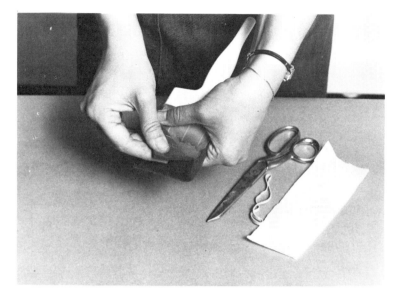

Tear off the sponge cover.

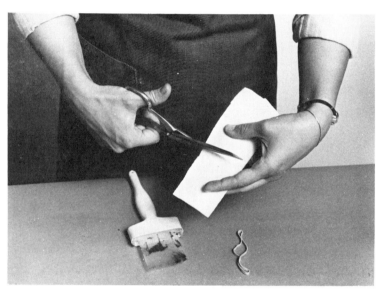

Cut a piece of felt to fit the flex-ible plastic tip.

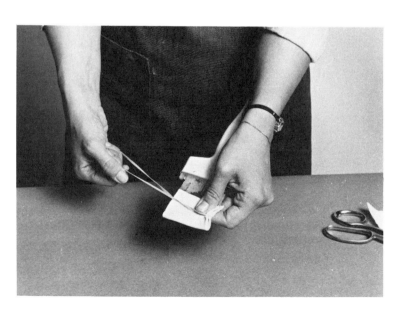

Fold the felt around the tip and secure it with a rubber band.

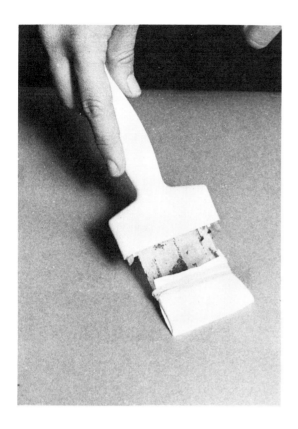

The brush is now ready to use for coating. The felt may be easily replaced for the next coating session.

from top to bottom. Smooth strokes and careful placement and removal of the brush at the edges almost always guarantee even application.

4. Dip the brush again and reapply the emulsion, this time moving the brush from left to right.

5. Some photographers wipe or buff the wet surface to pick up excess emulsion and to guarantee an even coating. Simply fold a piece of felt into a small, flat pad and briskly buff the surface—first left to right and then top to bottom. Discard this pad after use, because the emulsion will harden, and if the pad is reused it will scratch the paper's surface.

SENSITIZING THE PAPER

These emulsions have a low sensitivity to light. Consequently, some photographers choose to coat paper in dim room light. I recommend an ordinary yellow bug light (25 to 50 watts), placed about 5 to 8 ft. (1.5 to 2.4 m) away from the work area.

Formula I is to be used the same day as it is mixed and then discarded, so no additional mixing is required. Formula II must be mixed immediately before use in equal parts of A and B.

1. Pour into a shallow dish only the amount of blueprinting solution needed to coat the paper available.

2. Dip the homemade paddle brush or art brush in the solution. Squeeze off the excess liquid by pressing the brush against the side of the dish (see illustrations).

3. Starting at the top left of the paper, apply the emulsion evenly

Combine equal parts of Blueprint A and B solutions in a wide-mouth container.

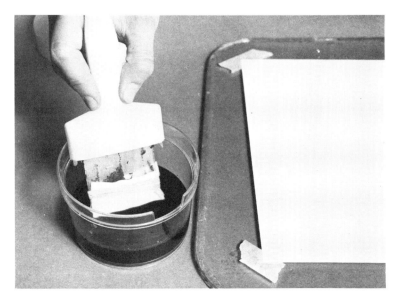

Dip the brush in solution. Allow it to sit in the solution for a few seconds so that the felt can soften and absorb the liquid.

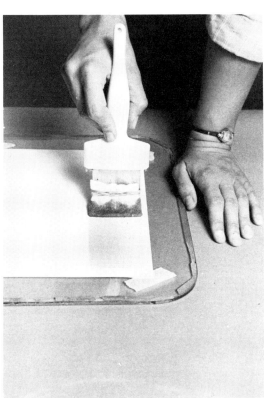

Apply pressure evenly when brushing on the emulsion.

Press the paddle brush against the side of the dish so that excess liquid drains off.

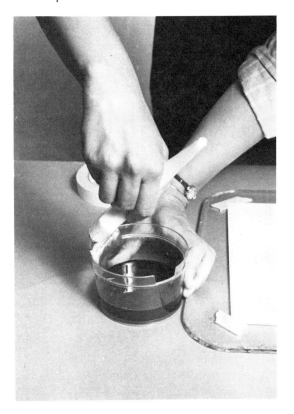

Apply a second application in the opposite direction from the first.

Buff off the excess with a clean piece of felt.

PAINTBRUSH APPLICATION

Application of blueprinting solution allows the opportunity to explore the painterly aspects of this and other similar nonsilver emulsions. You can choose where to apply the emulsion, so keep in mind the composition and overall layout of the transparency selected. Explore the possibilities of application by trying other tools such as a kitchen sponge, hair comb, or various ink pen points.

AIRBRUSHING

An airbrush can be used to apply the blueprint emulsion. Use a large painter's model or an inexpensive atomizer, available at most hardware stores. The airbrush method ensures an even coat of emulsion. Morever, it is a money-saving method because large amounts of emulsion are not absorbed into brushes.

If you choose to go the airbrush route, coat only as many sheets of paper as you can print in one session. Remember, sensitized papers *do not* keep overnight.

Caution: *Airbrushing should be done in a well-ventilated room and a mask should be worn over nose and mouth.*

DRYING PAPERS

Place the coated paper in a totally dark place to dry. Large, flat cardboard boxes are excellent drying bins for nonsilver papers, as are

Brush strokes create a loose painterly pattern.

A sponge applicator brush coats the paper heavily. Drying time is increased, but a deep blue is rendered in the finished print.

Spray on the emulsion using an inexpensive spray unit like the Preval unit shown in the photograph above.

Buff off the excess with a clean felt pad.

DETERMINING EXPOSURE

Sunlight is the fastest exposure source, but it is not always predictable; certainly it is not consistent. Other light sources (listed in the order of their intensity) are carbon arc lamps, quartz lights, unfiltered ultraviolet bulbs, sunlamps, and photofloods. If consistency is desired and you plan to do a large quantity of printing, I recommend building an ultraviolet light table as illustrated in Appendix B. If this is not feasible, sunlamps are your next best choice for controlling exposure.

Judging exposure by inspection is the simplest way of determining the correct intervals. The sensitized paper, when completely dry, is a yellowish green; as it is exposed to light, it shifts to a bluish gray. When this color is attained, exposure is completed and the paper should be developed immediately.

Of course, a test strip is a more reliable method of determining exposure. A test strip using 3-minute intervals is recommended as a starting point. For example, a sunlamp set 18 in. (46 cm) away will fully expose the blueprint in approximately 15 to 30 minutes, depending on the density of the transparency.

Although sunlight is not predictable, it does allow for relatively fast exposure times when all conditions are ideal. An exposure at noon can be as short as 5 minutes.

DEVELOPING THE BLUEPRINT

After exposure, the print can be developed. In yellow safelight conditions, immerse the exposed sheet of paper face down in a tray of running water for 5 minutes, or until the yellowish-green stain disappears. If staining remains in the highlight and unexposed areas after 5 minutes, continue development for an additional 2 to 5 minutes. After proper development, the image is blue green.

Note: If the yellow-green stain will not clear, prepare a tray of 1:3 Dektol solution and immerse the blueprint. Quickly remove it and rinse it in running water. Repeat this procedure until the stain disappears. If the staining still does not clear, the paper has been overexposed, or perhaps it is too absorbent. Therefore, it will have to be sized as discussed in Chapter 6.

WASHING AND INTENSIFYING

Thorough washing of the blueprint (which takes about 20 minutes) sufficiently stabilizes the image so that no further color shifts will occur. Some photographers claim that a few drops of hydrochloric

45

acid in a quart of water will fix the image, but I have found that this procedure is not necessary. Washing alone fixes the image. A hydrochloric acid solution, however, acts as an *intensifier* that enriches the cyan color inherent in this emulsion. With an eyedropper, measure two or three drops of hydrochloric acid per quart of water.

Caution: Hydrochloric acid is corrosive and poisonous. Wear rubber gloves. Do not inhale the fumes.

Immerse the completely developed blueprint in this weak hydrochloric acid solution until the desired value of blue is reached. Do not leave it in the tray longer than 3 minutes or the image may begin to reverse and fade. Wash the treated blueprint for about 10 minutes in vigorously running water.

If hydrochloric acid is not readily available, use household bleach diluted 1 oz. per quart (31 ml per liter) of water. Immerse the print in this homemade intensifier for 5 to

10 seconds, then quickly remove it. Wash it 10 to 20 minutes.

Hang the blueprint to dry. Drying can be accelerated in heated drying cabinets or with a handheld hair drier. This has no effect on the emulsion or the blue color.

OVERPRINTING

Success stories vary on the theory of overprinting—blueprint on top of blueprint. After the first emul-

A cyanotype print becomes three-dimensional with the addition of an embroidery hoop. Dan Vandevier printed this amusing scene on cotton cloth.

sion coat is dry and has been exposed, developed, washed, and dried again, the process can be repeated. Here, subjective decisions come into play. Does the desired result dictate blue-on-blue? If one layer is not intensified, the values of blue would be recognizibly different and would create an interesting effect. The concept should be considered in the exploration of this nonsilver process.

SENSITIZING FABRIC

Blueprint is an excellent emulsion for sensitizing fabric. The choice of fabrics is, of course, unlimited, but the best ones are the natural fibers such as muslin, 100 percent cotton, raw silk, and natural satins. Do not try to sensitize fabrics treated with water repellents such as Scotch Guard, because they will repel the liquid emulsion. A choice of col-

ored fabrics or preprinted patterns further extends the possibilities of experimentation with this process.

Painting the emulsion on fabric leaves a residue wherever the brush is placed; therefore, I recommend dipping or floating the fabric as follows:

1. Under yellow safelight, pour a large quantity of blueprinting solution into a tray (see illustrations).

2. Wearing rubber gloves, immerse

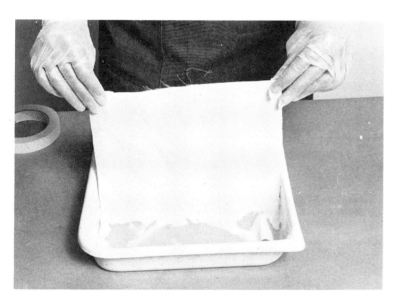

Lay the fabric in a tray filled with blueprint solution. Immerse the fabric completely.

Gently rock the tray until the fabric is completely saturated. Drain the excess liquid from the fabric and hang it in a dark place to dry.

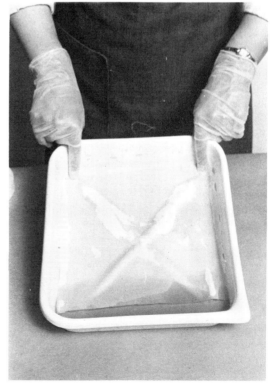

Elaine O'Neil created her Cow
Quilt by stitching together positive and negative cyanotype
prints. The 60- x 80-in. cotton
quilt is bordered in red and the
thread outlining the cows is red
also. A detail of the quilt shows
the embroidery more clearly.

the fabric in the solution, allowing it to float on top of the liquid.

3. Agitate the tray until the fabric is saturated.

4. Lift the soaked fabric by two corners, tilt it at an angle, and allow the excess liquid to drain into the tray.

5. Dry the sensitized fabric on a clothesline in a dark place. Be certain to use plastic clothespins. Put a tray under the dripping fabric to catch the excess emultion. Not only can it be recycled, but this emulsion stains most surfaces if it is not wiped up immediately.

Drying times vary, but fabric takes considerably longer than paper because its fibers absorb the liquid emulsion. When it is dry, contact-print on the fabric by any of the methods previously discussed for paper. A test strip is the best method for determining the correct exposure. Slight over-exposing guarantees that the emulsion has hardened into the fiber of

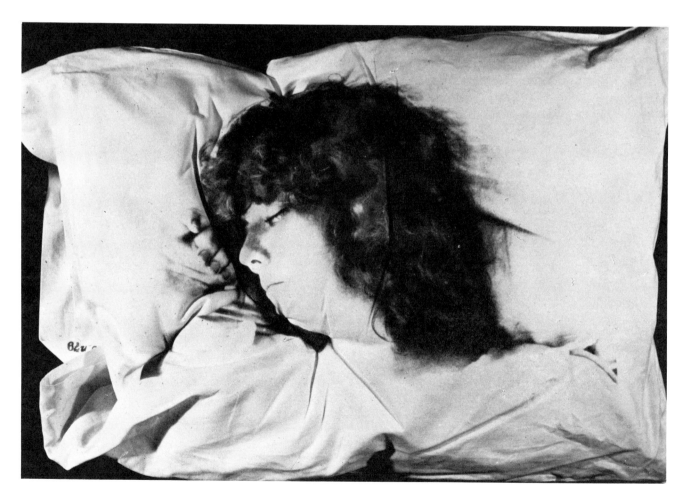

Elaine O'Neil blueprinted her Self Portrait Asleep *onto a pillow to create a life-size image.*

the fabric, but yellow staining is more prevalent with fabrics than with paper, so finding the correct time is imperative. You should continue to wear rubber gloves when developing the print in running water. Large amounts of unexposed solution are released at that point because fabric absorbs a greater amount of the emulsion than would paper.

Continue developing until the water running out of the tray is completely clear, with no blue tint. Again, the hydrochloric acid solution or the weak household bleach bath can be used to intensify the color. Wash the fabric slightly longer than you would paper after this step and hang it to dry. Drying can also be accelerated by the use of drying cabinets or a hair drier. The sensitized material may also be ironed.

TONING

The cyan color characteristic of blueprinting can be altered by chemical toning of the finished print. The resulting colors are unpredictable, and much experimentation is required to achieve any consistent results. Two widely used formulas are purple toning and red-brown toning. *Note:* For either of these toning procedures, working in a well-ventilated area is imperative.

PURPLE TONING

Bleach the print in a 5 percent solution of ammonia, rinse, and then immerse it in a tannic or gallic acid bath of 1 g to 100 ml of distilled water. Wash the print for 15 minutes. Do not overbleach, or the image will turn a dull brown when immersed in the tannic acid bath. There are no specific times for each step, and the desired color is achieved only by observation and quick reflexes.

RED-BROWN TONING

Chemically, the blueprint can be altered by toning to resemble a brownprint. This is encouraging considering the cost of silver nitrate, but unfortunately the unexposed areas shift after toning to a bright yellow. Red-brown toning gives the image an aged and vintage look.

Soak the print for about 5 minutes in a tannic acid solution of 6 g to 180 ml of distilled water. Then immerse it in a bath of sodium carbonate, 6 g to 180 ml of distilled water. Wash the print for 20 minutes.

Selectively toning certain areas of the finished print allows for a more creative extension of the process and should be considered a viable alternative to straight blueprinting.

Judy Miranda printed the negative for Death while the sensitized fabric's surface was still damp. The dampness of the emulsion caused the solarized look which is especially apparent in the lower left hand corner.

5 | Brownprinting

VANDYKE BROWN

Brownprinting is an old formula used primarily by draftsmen and engineers in the 1880s to reproduce drawings and mechanicals. The emulsion is faster than cyanotype, and the brown color shifts to a deep black through contact with heat.

Contemporary photographers revived this formula in the mid-1960s. Through workshops and college courses, the process has become just as well known as cyanotype in the area of alternative photography. The warm, rich tones of brown, the shorter exposure times due to the silver nitrate, and the altering of brown to black by heat make brownprinting an exciting process.

Unlike blueprinting, for which there are two formulas, brownprinting has one standard recipe. Vandyke brownprinting is not a true nonsilver process because it contains silver nitrate, which is an odorless, colorless, crystalline compound. It is also poisonous. When it comes in contact with skin, it burns and causes brown stains. Therefore, it is imperative to wear gloves and protective clothing when you mix the brownprint-ing formula. Although the other chemicals, ferric ammonium citrate (green scales or powder) and tartaric acid, are not so dangerous, their use also requires rubber gloves.

PREPARING THE CHEMICALS

This is the formula for brownprint-ing solution:

> 90 g ferric ammonium citrate
>   (green scales or powder)
> 15 g tartaric acid
> 37.5 g silver nitrate
> 32 oz. (.946 l) distilled water
>   at 85° F. (29° C.)

1. Under red safelight, measure all chemicals separately on a gram scale. Divide the quart of distilled water equally among four containers (see illustration on following page).
2. Dissolve each chemical separately in its own container of water. As you pour the compound, stir slowly with a plastic or glass rod. Be sure each compound dissolves

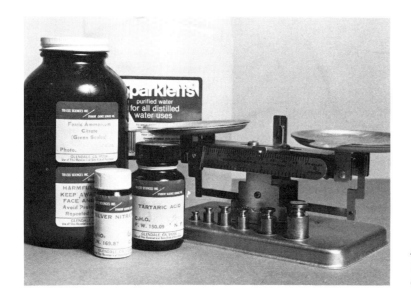

A gram scale is needed to measure the three brownprint chemicals.

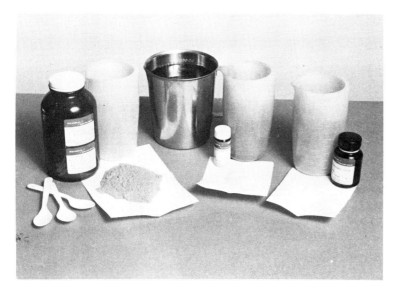

Use four containers to divide the water equally into four parts.

uniformly. When all solutions are saturated, you are ready to mix them together.

3. Combine the ferric ammonium citrate solution with the tartaric acid, then add the silver nitrate solution.

4. Thoroughly mix, and slowly add the remaining distilled water.

5. Pour the entire mixture into a brown glass or plastic bottle.

I recommend using a glass or plastic funnel when working with brownprinting solution because silver nitrate tends to react and eventually deposit onto other me-tallic surfaces. Clearly label the bottle "Contents Poisonous," and record the date it was mixed.

If stored in a well-stoppered brown bottle, away from direct light and intense heat, brownprinting solution has a shelf life of several months. Some settling occurs if it is unused for long periods; therefore, shake the contents of the bottle before you begin to work.

From this point on, the steps for brownprinting are almost identical to the steps for blueprinting presented in Chapter 4. Before sen-sitizing the paper to brownprint, read "Preparing to Blueprint" in Chapter 4. The same points about paper and coating tools also apply to the application of brownprint.

SENSITIZING PAPER

The next steps should be carried out under red safelight. Remember to wear rubber gloves when coating paper—this emulsion is poisonous!

1. Pour a small amount of brown-

After mixing each chemical separately in its container, combine all four to bring the solution to 32 oz. and pour it into a brown bottle. Date the contents.

printing solution into a glass or plastic bowl.

2. Stir and dip the brush or paddle into the mixture and apply it to the paper. It is best to start at the top and make long, even strokes to the bottom. (Refer to the illustrations in Chapter 4 under "Sensitizing the Paper.")

3. Repeat this application method from side to side, thoroughly covering the paper.

4. You may buff the wet surface to remove excess emulsion, as described in Chapter 4. You might explore some of the various other ways of application.

Caution: Do not airbrush this emulsion. Brownprint, when atomized, creates toxic fumes.

Sensitize only as many pieces of paper as you will need for a printing session, because, as in blueprinting, they will not keep overnight. Dry coated papers in a dark place. Brownprint is more sensitive than blueprint to fogging in the wet state. When the paper is dry, usually in 20 to 30 minutes, place the transparency in contact with the sensitized paper. You are now ready to print.

EXPOSURE AND DEVELOPMENT

Selecting the proper light source for your needs is discussed in Chapter 4; the same principles apply to brownprinting.

The brownprint emulsion is twice as fast as the blueprint; therefore, exposure times are approximately half as long. Again, a test strip is the best way to determine correct exposure times, but instead of 3-minute intervals, use 1 or 2 minutes.

55

To judge a brownprint by the inspection method, look for the emulsion to shift from a yellowish orange to a red brown and then a deep orange brown as it hardens in the clear areas of the transparency. When a deep brown color appears, the print is ready to be developed.

DEVELOPING THE BROWNPRINT

Under red safelight, immerse the exposed sheet of paper face down in a tray of briskly running water. After a preliminary development of 5 minutes, the room lights may be turned on so that you can examine the image. If a light brown stain remains in the highlights and other unexposed areas, double the developing time to 10 minutes. As in blueprinting, it is impossible to clear stains caused by over-exposure. After 20 minutes of developing, if residue is still visible, consider sizing the paper or revising the exposure times to eliminate this problem. After development is complete, the image will be an orange-brown color. Some photographers prefer this color and simply give their prints a final wash for 20 minutes, hang the print to dry, and consider the process complete.

FIXING

It is my experience that because brownprinting is a silver process, the silver nitrate salts must be fixed or the image will continue to shift in color with further exposure to light. Two fixing bath formulas are available. One is a simple solution of one part standard paper fix to twenty parts water. The other

formula is 1 tablespoon sodium thiosulfate (the chemical compound used in fixer, without the hardener) per quart of water. Both work successfully, and both turn the image a deep vandyke brown color from which the process gets the name.

To fix, simply immerse the developed print in the fixing bath and agitate it until the desired brown is achieved. This fixing tends to clear the highlights and in some cases can save an overexposed print. Do not overfix, because the image can be bleached away. Discard the fixer after use.

Wash the print for 15 to 30 minutes, depending on the circulation of the running water over the surface of the print. A hypo clearing agent will cut washing time in half. Hang the print to dry. You *cannot* accelerate the drying time by the use of heat. Air-dry only.

If, after drying, the image fades in intensity, the print has not been washed sufficiently. There is no way to save an underwashed print; therefore, if you are in doubt, overdoing the washing is always good insurance. Brownprinted images, if correctly done and preserved with the same care as silver prints, will last as long as the paper on which they are printed.

APPLYING HEAT

When a brownprint is completely dry, a black image can be obtained by applying heat to it. In a dry-mount press heated to approximately 300° F. (149° C.), cover the print with a sheet of lightweight bond paper to protect it from sticking to the plate and press for 10 to 15 seconds. Repeat if necessary,

until the print is a deep, rich black. If a dry-mount press is not available, cover the print with a piece of bond or lightweight paper. Heat a household iron to the cotton setting. Iron the protected print and lift the cover sheet to check for the desired black. This altering from brown to black is especially helpful when you want a full black-and-white tonal range on other than commercial silver paper.

COMBINATION PRINTING

Brownprint and blueprint can be combined on the same piece of paper, but doing so is a bit tricky. Blue and brown cancel each other out because of their different chemical components, and overprinting is not possible. Applying them side by side, printing first blue and then brown, has been done. Over-printing of brown-on-brown has never been successfully achieved, because the fixing of the second application tends to reduce the intensity of the first printing.

SENSITIZING FABRIC

The sensitizing of fabrics with the brownprint solution is very similar to that of blueprint. Application with a paintbrush creates an uneven and blotchy surface, so all fabrics must be dipped.

Follow the step-by-step procedure outlined in Chapter 4, "Sensitizing Fabric." Prior to dipping, be sure the brownprint solution has been thoroughly mixed so that the silver nitrate is suspended evenly. As discussed earlier, it is imperative to wear rubber gloves when dipping and developing sensitized fabrics.

Because of the excessive saturation of the fabric, exposure times should be increased 10 to 15 percent for fabrics. Wash the fabric thoroughly, increasing this time by 10 to 15 percent as well, so that staining will not occur after the fabric has dried. Do not iron the fabric unless you want the color to be black. Stretch the fabric on a wooden frame, or staple it to a board to prevent wrinkles during drying.

Refer to Chapter 12 for additional ideas for working with fabrics and for various finishing techniques.

Photograms of leaves and gloves were juxtaposed with a photographic negative to create this image. This untitled brownprint by Darryl Curran is included in the collection of the Museum of Modern Art in New York City.

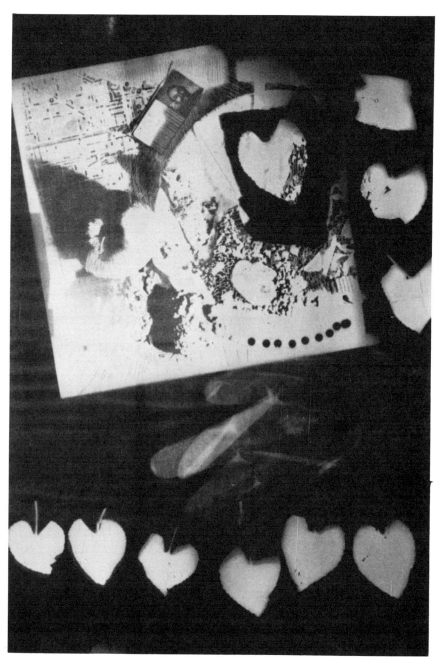

Todd Walker used green fabric as a base for his gum print of a nude figure. The model's skin tones are true, while the fabric she is wrapped in picks up the natural hue of the base.

6

Introduction to a Full-Color Palette

GUM BICHROMATE

Unlike blueprinting and brownprinting, where one is limited to a monochromatic emulsion, gum printing introduces the nonsilver photographer to a full-color palette. An English photographer, John Pouncy, invented and patented this process in 1858, and it became quite popular near the end of the nineteenth century when such photographers as Robert Demachy and Alfred Maskell began to work in the medium. At the same time, American photographer Edward Steichen executed several multiple-color gum prints. He also combined gum with other nonsilver processes such as cyanotype. Steichen's full-color gum prints were reproduced in Alfred Stieglitz's *Camera Work* and are considered by many as the starting point for discussions of the history of color in photography.

Gum printing, also called the gum bichromate process, uses gum arabic with a light sensitizer (usually ammonium dichromate) and ordinary watercolor paints as the pigments. *Varied* is the word that best describes this process and its results. The list of variables in gum printing could well fill an entire book. The formula is standard, but the procedures are numerous for mixing, sizing the paper, applying the emulsion, developing, and recoating.

Patience is essential when attempting a gum print. The failure rate is high because the emulsion is delicate. Moreover, staining caused by multiple-color printing and reregistration of more than one negative are also possible problems.

The variables will be discussed throughout this chapter, so that if one procedure does not bring the desired result, an alternative method can be tried. Keeping a notebook and recording each step will aid you in tracing problems that might arise with gum printing.

PREPARING TO GUM PRINT

The selection of the paper is critical because it must be able to withstand long periods of soaking and drying and drying again. The development times for gum printing are anywhere from 30 minutes to 1½ hours per color application; there-

Spray starch directly onto preshrunk dry paper as a quick method for sizing.

fore, 100 percent rag fiber papers such as Rives BFK (smooth), Arches (textured), or any other fine, high-quality watercolor paper will suffice. Smooth-surfaced papers are best for generating fine details. A paper with a tooth to its surface, however, renders a more characteristic gum print and is recommended for multiple-color applications.

These papers are quite expensive, so to save money, consider an excellent 25 percent rag paper made by Lennox called Art Print. There is enough rag fiber in this smooth-surfaced paper to withstand the long immersion in water.

If you plan to do multiple-color printings, it is imperative to preshrink the paper before sizing. Soak the paper for 15 minutes in a tray of 100° F. (38° C.) water. Hang the paper to dry, slightly bowed, so that it can shrink easily.

SIZING THE PAPER

When the paper is completely dry from preshrinking, it is ready to be sized. Sizing means coating the paper with a substance that prevents staining in the unexposed areas. During multiple-color printing, the residue left by each color muddies the highlights. Sizing usually helps retain clean highlight areas.

There are four common sizing formulas, each with special characteristics and advantages, as well as disadvantages. The simplest uses ordinary spray starch. Any commercial brand will work, but I recommend Miracle or Faultless. The starch should be sprayed directly

onto the surface of the preshrunk paper. Start at the top and spray in horizontal strokes, working from left to right. Allow this coat to dry completely. Then apply the starch again, working with vertical strokes from top left to bottom right. Let this dry, then repeat the entire procedure (see illustration). This method is excellent for two or three colors but will not endure more than three coats before either the starch lifts off or staining occurs. (*Note*: You might consider reapplying a thin layer of spray starch between each color application.)

A second method requires the use of Elmer's glue (or a similar brand) diluted 1:4 with water. Fill a tray with this sizing solution. Immerse the paper and turn it over two or three times, allowing it to absorb the solution. After 15 minutes, remove and squeegee the paper to remove the excess. Allow it to dry. Repeat the procedure if more than two colors are to be printed.

Liquitex Gesso, diluted to brushing consistency with warm water, is a third sizing formula. Brush it directly on the paper and hang it to dry. The same steps apply with Gesso as with Elmer's glue; recoating is suggested for multiple-

color printing. Gesso diluted 1:3 with warm water is also an excellent sizing for fabric. (*Note*: Gesso must be hardened with the same solution of formaldehyde that is required for unflavored gelatin sizing when working with fabric.)

The most durable sizing is unflavored gelatin hardened with formaldehyde. This method is the one most often used with multiple-color printing, for it can withstand anywhere from three to more than twelve coats of gum emulsions.

Add three packages of unflavored gelatin, approximately 21 g, to a quart of room-temperature water. Allow the gelatin to swell (about 10 minutes), stirring periodically to prevent hardening. Warm this swollen solution to about 100 to 110° F. (38 to 43° C.). Immediately pour the liquid into a tray and immerse the paper. Several sheets can be immersed at once, but it is important to shuffle them continually so that each can absorb the liquid.

Remove each sheet of paper separately, allow the excess to drip back into the tray, and squeegee the paper between two glass rods. This may require the help of a friend to hold the glass rods while you pull the wet paper through. The purpose of this procedure is to

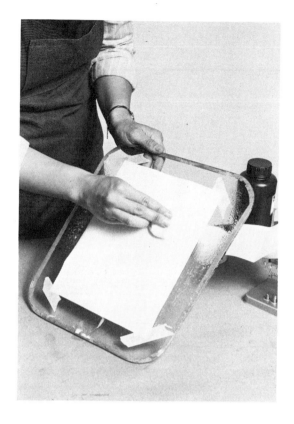

Puddling of the starch may occur, so buff the surface with a piece of cotton or felt to pick up the excess.

remove any air bubbles that might be present.

Jack Butler, an artist who works with gum printing, built a squeegee that works on the same principle as a pair of glass rods. He uses wooden dowels, opposite each other and slightly off-register, assembled parallel in a U-shaped stand. The gelatin-soaked paper can be easily pulled through this apparatus and then hung to dry. (Appendix C gives simple instructions on how to make this roller squeegee.)

When completely dry, the gelatin-sized surface must be hardened with 37 percent formaldehyde diluted with water. Mix 25 ml of formaldehyde with a quart of water.

Warning: Do this hardening process in a well-ventilated room. Inhaling the fumes could be harmful and even fatal.

Pour the formaldehyde into a tray. Immerse and shuffle several sheets, making sure the solution covers and completely soaks into the paper. Remove, drain, squeegee, and hang to dry.

This fourth sizing method is excellent for all gum-printing needs. It works best for solving staining problems should they occur with the blueprint and brownprint emulsions. (*Note:* Because of the short shelf life and the fumes with this sizing method, it is best to size many sheets of paper at one time. The unflavored gelatin/formaldehyde sizing will coat approximately twenty-five 11 × 14 sheets of paper.)

This sizing solution can be saved by refrigeration in a tightly covered glass jar. It will keep for several days and can be reheated for use, but keep in mind that repeated heating of the gelatin lessens its ability to harden properly.

THE CHEMICALS

Gum arabic is the adhesive substance in which the pigment and sensitizer are suspended prior to exposure. It is available as a light tan syrupy solution, in liquid form, or in a crystal or "tears" state to be mixed with water. Liquid gum arabic may be purchased from lithography supply stores or chemical houses as 14-degree Baumé, and some suppliers now label it "photographic" specifically for this printing process. Gum arabic has a tendency to sour unless it is refrigerated, but 14-degree Baumé liquid gum contains carbolic acid (phenol) to prevent souring. This acid prevents bacteria from growing in the solution and promotes a uniform consistency of the liquid.

Liquid 14-degree Baumé solution simplifies coating the paper, but if it is unavailable, you can mix your own from "tears" of gum. Measure 90 g of gum arabic onto three or four folded layers of fine cheesecloth. Pull up the corners, twist the cheesecloth into a hard, round bag, and tie it securely. Suspend the bag in about 280 ml (a 1:3 ratio) of room-temperature water, below water level, and allow it to sit for approximately 24 hours. Through this suspension method the cheesecloth collects any impurities in the gum and should create the correct viscosity or saturation of liquid gum arabic needed for gum printing. *Note:* A specific gravity reading with a hydrometer to determine if the dissolved gum arabic is 14-degree Baumé is also required. A hydrometer can be purchased at a scientific supply store.

This method of mixing results in an odor that is quite offensive. Covering the solution with plastic wrap will help. Use several drops of formalin to help curb the odor as

The chemicals and pigments shown in this photograph are the basic supplies needed to gum print.

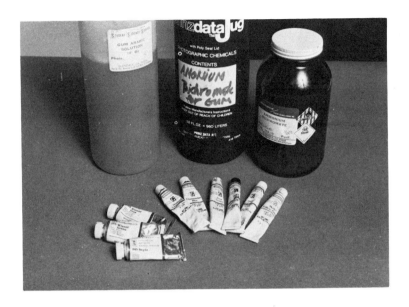

well as to preserve the gum from any further souring. Refrigerate the dissolved solution in a covered jar. It can be used for up to two weeks.

The second most commonly used chemical, because of its greater sensitivity to light, is ammonium dichromate or bichromate (the names are used interchangeably), an odorless, red-orange crystalline chemical that must be diluted with distilled water to make a working solution for gum printing.

Warning: Each of these chemicals is harmful and poisonous. Ammonium dichromate on contact with skin can cause severe irritation, or even burns. If this occurs, flush the skin with running water. Use rubber gloves. If you accidentally ingest it, consult a physician immediately.

THE PIGMENTS

The pigments used for gum printing are usually tube watercolors. Pigments may also be purchased in the crystalline state, to be ground or crushed with a mortar and pestle, but tube watercolors are simpler to use and allow for more accurate measuring. The finer the quality of watercolor paint, the purer the pigment composition. Excellent choices are Winsor & Newton, Grumbacher, Pelikan, and Shiva, which can be purchased a tube at a time or in kits. I recommend purchasing a kit of starter colors; yellow, red, green, blue, black, violet, and orange are a good palette with which to begin. These

paints can be mixed to create new colors. It is not necessary to work exclusively with tube colors.

An additional chemical is potassium alum or aluminum potassium sulfate (APS), the use of which is optional. This crystalline compound is mixed with warm water to make a post-printing hardening bath for the swollen gum emulsion. It also clears the yellow-orange staining sometimes caused by overexposure.

THE BASIC FORMULAS

Under red safelight conditions, prepare the following three parts (see illustration):

Part A:
Gum arabic—14-degree Baumé solution is the proper working stock solution.

Part B:
Ammonium dichromate—Mix 25 g with 1,000 ml water at 100° F. (38° C.) to make a stock solution. Make certain the crystals are suspended evenly in the water. This saturated dichromate solution has an indefinite shelf life and does not need refrigeration. Store it in a brown bottle.

Part C:
Pigments—Using tube watercolors, measure out squeezed links of the desired amount of pigment. A 1-in. (25-mm) link is for a light color saturation, a 2-in. (51-mm) strip is for medium color density, and a 4-in. (102-mm) tube link will create a heavy saturation of color. A piece of glass with a strip of tape attached can indicate 1-in. (25-mm) increments conveniently (see illustration).

Suspended in the gum are ammonium dichromate and a pigment. When exposed to ultraviolet light, the ammonium dichromate reacts with the gum and the pigment, causing the exposed areas to harden and become insoluble. They attach to the paper sizing and to the paper's surface as well. Areas that receive no ultraviolet light (highlights) drop off when developed. Middle tonal areas retain varying layers of soluble and insoluble (hardened) emulsion, depending on the densities of the negative. These middle areas can be controlled and further altered during development as discussed later. Shadow areas harden completely, attaching to the paper.

The variables in mixing a gum-print emulsion are infinite. The ra-

Tape a small ruler or a piece of masking tape with one inch intervals marked on it onto a piece of glass. Squeeze out pigment beside inch mark.

Immerse the spoonful of pigment into the gum arabic and stir vigorously until the solution is thoroughly mixed.

tios of gum arabic to ammonium dichromate to pigment can be mixed in an endless number of combinations.

Here, again, the need for keeping data is imperative if you ultimately wish to control this process. Limiting yourself to one paper, such as Rives BFK, one light source, such as a sunlamp set 3 ft. (1 m) from the contact frame, and one contact negative is suggested to help you set up a starting point. Next, select one formula of combining the gum arabic with the ammonium dichromate solution. In the beginning, the amount of pigment used should be the *only* variable.

The are two standard formulas for combining gum arabic (Part A) with ammonium dichromate (Part B) and pigment (Part C).

Formula I is a 1:1 solution of A and B with varying amounts of pigment. Under *red* safelight, measure the desired amount of pigment and combine it with 30 ml (1 oz.) gum arabic (Part A) in a plastic cup or glass bowl (see illustration). Stir until the pigment is completely saturated in the solution. Make sure that none of the watercolor paints have settled on the bottom or sides of cup or bowl. Add 30 ml (1 oz.) of ammonium dichromate

Add ammonium dichromate and stir again until all three are combined uniformly.

Pour the gum print solution into a wide-mouth container for easy access with the coating tool.

mended for use with halftone negatives. This formula is also excellent for multiple printing of more than one color and allows for build up of emulsion layers.

SENSITIZING THE PAPER

There are two procedures for coating or sensitizing paper. One method utilizes the paddle brush described in Chapter 4; the other requires two soft camel's hair or sable paintbrushes.

The homemade paddle brush is easier to use and produces a more even coating. The felt fabric on the tip of the flexible plastic paddle soaks up just enough gum solution so that there are no drips or puddles of emulsions. Spreading is smoother, and the emulsion does not usually oversaturate the paper. Blending with a clean piece of felt completes the application (see illustration).

The sensitizing procedure is done under red safelight. Using the paddle brush, start at the upper left and apply the emulsion in vertical strokes from top to bottom. Dip the brush again and drain off the excess emulsion against the side of the dish. Quickly apply the emulsion in horizontal strokes from upper left to upper right. Before the emulsion can begin to dry, blend or buff the paper with a clean felt pad, working first top to bottom and then left to right. Place the pa-

stock solution (Part B). Stir until the mixture is uniformly saturated. The color of the mixture will shift because of the reddish-orange color of the ammonium dichromate. This may throw you off visually when you coat the paper, but the "discoloration" washes off during development. Formula I is better for single-color gum prints, because it renders strong contrasts in the shadows, holds up well in the middle tones, and usually clears the highlights easily. Staining can occur, but it is caused by improper sizing or too heavy-handed an application of gum-print emulsion.

This 1:1 solution is excellent for heavy saturation of pigments (4-in. or 102-mm strip) because of the extra gum arabic.

Formula II is a 1:2 solution of A and B with varying amounts of pigment. Follow the same procedure as for Formula I, *except* use 15 ml (½ oz.) gum arabic (Part A) with the desired amount of pigment. Then add 30 ml (1 oz.) ammonium dichromate (Part B) and mix thoroughly as discussed above. This formula thins the coating mixture and shortens the exposure times. Formula II softens contrasts in the resulting print and is recom-

Jack Butler applied Dr. Martin's Dyes to a cyanotype to obtain the full-color image shown above.

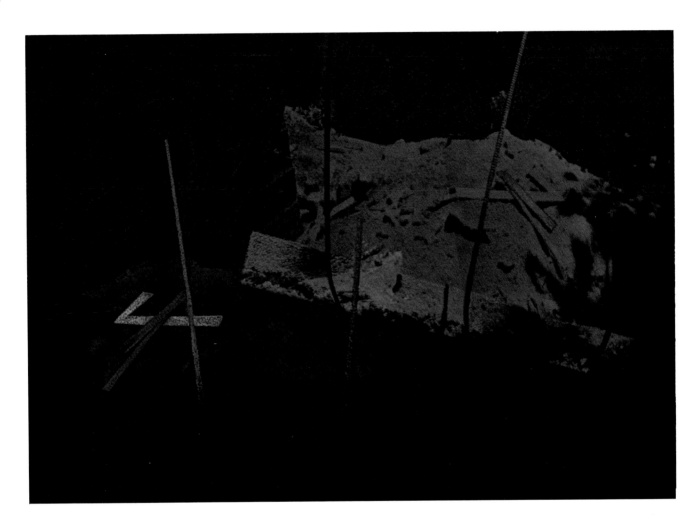

Olga Seem used colored pencil on a brownprint to create Construction Site #1. The brownprint's base is watercolor paper. Courtesy Eve Mannes Gallery, Atlanta, Georgia.

Like Construction Site #1, Construction Site #2 *incorporates hand coloring over a brownprint on watercolor paper. Courtesy Eve Mannes Gallery, Atlanta, Georgia.*

Todd Walker created this lovely soft-focus image by printing multiple layers of gum bichromate.

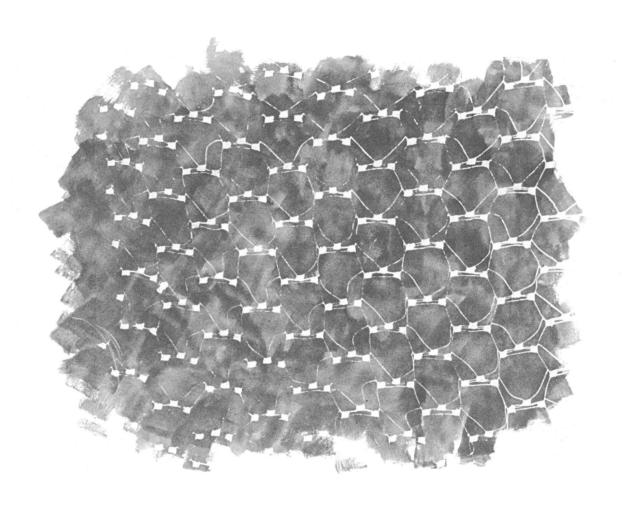

Anne Flaten Pixley experimented with natural pigments such as Chianti and grape juice to obtain this subtly toned gum print. Red Net #2 (shown here) is a photogram.

Jim Cokas's construction
entitled We Played Cards
'til Our Decks Were Worn
Out incorporates conven-
tional silver prints and cyanotype on silk. The bits of
playing cards in this
viewer-participation piece
are real.

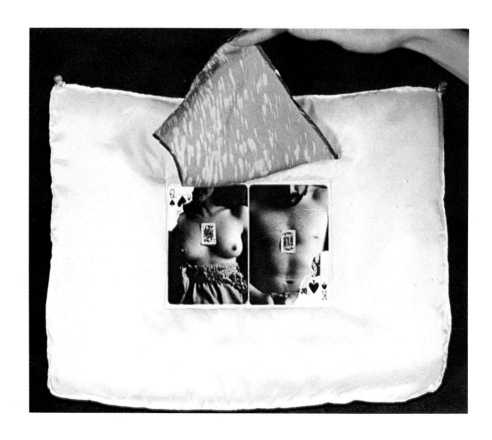

Elaine O'Neil's series of stuffed postcards are cyanotypes embellished with paint, embroidery, rubber stamps, and color Xeroxes. They measure 3½- x 5½-in., and have actually been mailed. Shown above is the front and back of the postcard entitled Smoky Mountains National Park.

Todd Walker generated the print entitled Edie *by using layers of gum emulsion in conjunction with color separation negatives.*

Apply the solution as you would blue- or brownprint, coating the paper first in one direction and then in the other.

Buff off the excess to create an even surface application. Use a felt fabric pad.

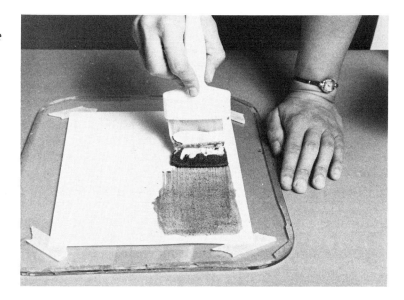

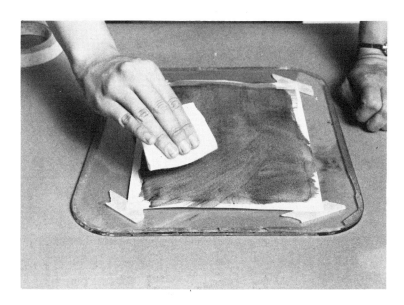

per in a totally dark place to dry. Discard all felts and cover the paddle brush with fresh fabric for the next color application.

The second method of application entails using two paintbrushes about 2 in. (51 mm) wide. One should be designated a "wet" brush for emulsion application and the other a "dry" brush for blending.

Dip the "wet" brush in gum emulsion, fully soaking the fibers. Drain off the excess by pressing the brush against the side of the container, and briskly brush the paper in a vertical direction. Watch for drips caused by an oversaturated brush; staining can occur from heavy brushwork.

Reapply the solution, working horizontally. Now quickly rework the wet surface with the dry brush, blending any streaks first vertically and then horizontally. There should be light "feathering" of brush marks on the edges of the paper from this method. Place it in a totally dark place to dry. Wash both the wet and dry brushes thoroughly to get them ready for application of a new color. Having a few extra "dry" brushes around precludes having to force-dry the brushes with a hair drier to expedite the coating procedure between color applications.

Dry coated papers in total darkness.

Coat only as much paper as you plan to expose in a single printing session, because sensitized papers

will *not* last overnight. Gum prints should be exposed as soon as they are completely dry, because gum arabic hardens if it is allowed to sit for long periods. Develop them immediately upon completion of exposure.

Discard all color gum mixtures. Once A and B have been mixed, the solution cannot be stored for use at another time.

EXPOSURE

The most successful gum-printing results are usually from high-con-

trast negatives, but continuous-tone litho negatives will work. For these, however, exposure and development are more critical.

Place the sensitized paper and negative in a contact frame. Expose them to the chosen ultraviolet light source—sunlight, carbon arc, sunlamp, quartz, or photoflood—and make a test strip or test print at 1-minute intervals. Many photographers believe direct light causes more contrast. Try both direct and diffuse light. Comparison is the only way to determine which light source provides the desired result.

Exposure times vary from 3 to 30

minutes, depending on the light source. For example, the exposure time under a photoflood 3 ft. (1 m) from the contact frame runs 3 to 8 minutes. For a sunlamp 3 ft. (1 m) away, the times are 8 to 12 minutes, and in bright sun, the exposure time is 3 to 6 minutes.

A high-contrast negative requires less time for complete exposure of the shadow areas, whereas exposure of a continuous-tone negative must be judged primarily by the middle tonal areas. Again, a test strip for each negative is the best bet.

Eposure times also vary depending on the amount of pigment used. A heavy saturation of pigment requires a longer exposure than a lighter mixture. Lighter colors print more quickly than darker colors, and the blue end of the spectrum is more sensitive than the red. Remember also that Formula II is more sensitive to exposure than Formula I.

Keeping a notebook of exposure times is a must for each negative, gum-printing formula, pigment amount, light source, and paper stock. This information will allow you to duplicate images later as well as overprint different color emulsions without having to make a test strip each time.

DEVELOPING THE GUM PRINT

This is a simple water wash-off development process, but unlike in blue- and brownprinting, the gum emulsion swells when immersed in water, so the print must be handled very carefully.

There are two slightly different procedures for gum-print development. One works on the principle of minimal agitation during development; the other encourages a more aggressive handling and agitation of the gum print.

Developing is merely the floating away or washing off of the soluble gum in water. The first procedure purports that a correctly exposed gum print will develop itself if it is left standing in a tray of room-temperature water, face down, from 30 minutes to 1 hour. If the image develops in less than half an hour, it is underexposed; if an image shows up very quickly, it will eventually wash off completely and the print will be lost.

This form of development is time-consuming, but it is the one most recommended. It should be used with Formula I because the 1:1 relationship of gum arabic to ammonium dichromate can sometimes cause flaking if the print is agitated too vigorously.

The other procedure involves less time in the development stage. It has proven especially successful for developing large numbers of prints, and it works best with Formula II.

1. In ordinary room light, set up two trays of room-temperature water. Set up a third tray with cold, slow-running water; the water should change completely every 15 minutes.

2. Immerse the exposed gum print face up in the first tray. Gently rock the tray back and forth until the reddish-orange ammonium dichromate begins to wash off the print and discolor the water.

3. After approximately 5 minutes, remove the print carefully from this tray and allow the ammonium dichromate and unexposed pigment to run off one end.

4. Place the print with its swollen emulsion face down in the second tray. Agitate the tray slightly, observing further ammonium dichromate and pigment discoloration from this wash-off development.

5. Let the print stand for about 20 to 30 minutes; then place it in the third tray of slow-running cold water, first for about 5 minutes face down and then for another 10 minutes face up.

6. After this series of steps is completed, lift the print from the running-water bath by its corners, and if no further ammonium dichromate or color runs off, development is complete.

7. Return the print to the third tray if color runoff continues, and wash until the print drips clean water.

8. Lay the completely developed print flat and face up on a glass surface for about 20 minutes, until the image has begun to set up.

9. Hang it to dry.

If you plan to print multiple colors, it is best to dry the prints on aluminum window screens, so that there is no stretching caused by hanging. Stretching can throw registration off between color applications.

FORCED DEVELOPMENT OR "WORKING" THE EMULSION

With either form of development, the emulsion of gum can be manipulated, to force development of the highlight areas and overexposed shadows or to alter middle tonal values.

Halfway through development, the print may be turned face up in the water and the emulsion manually removed or "worked" by several different methods. The most obvious is to rub an area with a finger or to scratch away at the emulsion with a thumbnail. This works for alterations of some areas, but for selective removal of small areas of emulsion a fine paintbrush, cotton swab, or pointed instrument works best.

To soften and lift large areas, a fine spray of water from a hose or squirted under pressure from an atomizer will remove the emulsion without leaving marks, which utensils can cause. A large, soft paintbrush works, but consider that it will leave marks as well.

OPTIONAL CLEARING AND HARDENING BATH

Development by either method should successfully clear any orange stain caused by the ammonium dichromate. If it still remains, it can be removed at the end of development by immersion in a 5 percent alum bath. Some photographers also use this to harden the swollen gum emulsion, especially between coatings of multiple prints.

There are two formulas for this clearing and hardening bath. Either renders successful results.

Formula A
   25 g potassium alum
1000 ml water

Formula B
   15 g aluminum potassium
         sulfate (APS)
300 ml water

After immersion in an alum bath for 5 minutes, the gum print must be washed in a tray of water at room temperature. Change the water two or three times over a period of about 15 minutes. (*Note:* you can use a 5 percent solution of potassium metabisulfite or sodium bisulfite, because an alum bath does not promote print perma-

nency. I have not noticed any color shifts from the use of an alum bath, however.)

REGISTRATION FOR MULTIPLE PRINTING

Success in multiple printing requires careful preparations before the actual printing can begin. Preshrinking and sizing with unflavored gelatin and formaldehyde are the first basic requirements. The second step is the selection of a system for negative registration.

The most common registration method uses pushpins and negatives measuring exactly the same size. Begin by placing the negatives all in register on the sensitized paper. Insert two pushpins about 1 in. (25 mm) on either side of one corner of the negatives. Then place two more pushpins in the opposite or diagonal corner. Tape the bottom negative in place. Remove the pins and print in a contact frame. After exposure and development, when the print paper is completely dry, coat with a second color. When that coating is dry, replace the four pins in their previously marked holes, and slip the same or a second same-size negative in place. Secure with tape. Expose, process, dry, and recoat with a second color.

A second registration procedure requires masking the negatives with commercial photo-masking paper, or simply taping all four sides with black tape. Register all negatives on a light table, and secure them with masking tape one on top of another by the edges. Punch all the masked negatives together in each of the four corners using an ordinary hole punch. Registration of each color is then ac-

complished by simply lining up each dot printed on the border.

A third method calls for outlining the first negative prior to printing, using either a soft-lead pencil or a ballpoint pen. When registering the second negative, place the sensitized paper on a low-wattage (about a 50-watt bulb) light table, and put the first negative within the outline previously drawn. Tape along one edge. Place the second negative on top and align it with the first. When the negatives are in registration, tape the second negative to print on an opposite edge. Lift the second negative and carefully remove the first. Tape the other side of the second negative and print. Registration of a third and all successive negatives should be done by lining them up with the first negative. It may be necessary to outline the initial registration marks again, as the layers of gum emulsion begin to build up on the surface of the print.

PRINTING ON FABRIC

Gum printing on fabrics is a difficult and delicate procedure, and the failure rate is quite high. Keep detailed notes, and work through the problems one at a time. You can obtain satisfactory and interesting results, but you need patience and much trial and error.

Preshrink natural-fiber fabrics—cotton, muslin, and so on—and size fabric with Gesso diluted 1:3 with water and hardened with formaldehyde.

Mix Formula II and coat the fabric with a large brush. When it is completely dry, expose it in a contact frame and increase the exposure time by 10 percent over that of paper.

Develop using the method of less agitation. Allow the fabric to float in standing water for a minimum of 30 minutes. Then remove it to a second tray and increase the water flow. Force-develop by rubbing the highlights or unexposed areas to clear any staining. Continue development until the desired results are achieved, then wash the finished fabric in a tray of running water for 20 to 30 minutes. Drain and lay the fabric flat to dry. In this way, swollen emulsion can harden into the fabric's fibers. Blotting with paper towels absorbs excess water and speeds the drying time.

Note: If the image releases from the fabric and floats away, increase the exposure time. If the image is overexposed, continue development, aggressively rubbing the dense areas.

Use of the clearing and hardening bath is optional. Use the hole-punch method for registering more than one negative. Ironing the gum print on the nonemulsion-side of the fabric using a low heat setting will flatten it and aid in correctly registering the negatives for multiple-color printing.

Gum printing is a beautiful process. The inherent nature of the medium, the variables in coating procedures, the full-color palette, the degrees of surface development, and the number of color coats applied—all these make gum printing as exciting as when John Pouncy patented the process more than 100 years ago.

7

A Commercial Proofing Process

KWIK-PRINT

Photographers, especially those working with nonsilver processes, are noted for discovering unusual applications of commercial printing processes. Kwik-Print is a good example of this creative crossover. Originally used by printers to proof full-color separations, it is now available to nonsilver photographers through Light Impressions Corp.

Kwik-Print is an alternative to gum printing. Like gum printing, it is a contact-printing process requiring same-size negatives or positives. Unlike gum printing, there is a low failure rate with Kwik-Print because many of the problems characteristic of gum, such as staining, shrinkage of paper, and vulnerability of the swollen emulsion, do not exist. Kwik-Print is a sturdy emulsion with which to work. Multiple-color printing is accomplished quickly and easily because inherent in the colors are driers that speed up the coating process. No sizing or preshrinking is required if the vinyl sheets supplied by the distributor are used. A reducer is available to salvage overexposed prints, and all of the manipulation or force-developing techniques of gum, such as rubbing away unexposed areas (see Chapter 6), can be executed without any damage to the Kwik-Print.

BASIC MATERIALS

Basic supplies needed for this process include the following.

Stable Base Sheets. These are vinyl sheets ranging in size from 8 × 10 to 54 × 75. There are three types of surfaces from which to choose: Hi-Con V for line negatives, Wide Tone P and V are continuous-tone negatives. They are *not* light-sensitive and will last indefinitely. Each sheet is notched so that both the type and working side are easily identified. The notching is on the short side. *Note:* Fine artist's papers with rag fiber content can also be used, but they must be sized (see Chapter 6) with the unflavored gelatin/formaldehyde recipe before coating. Fabrics can also be coated. This procedure is discussed later.

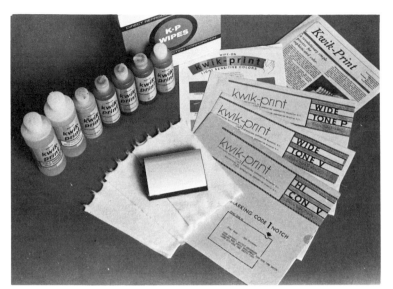

A kit can be an advantage when you are first starting to Kwik-Print. The photograph at left shows the Fine Arts Advanced Kit.

KWIK-PRINT COLORS:

There are fifteen light-sensitive, commercially mixed colors available. They include lemon yellow (process color), magenta (process color), medium blue (process color), black (process color), crimson, pink, geranium red, flesh tint, pale yellow, ultramarine blue, pale blue, gray, light emerald, dark brown, and opaque white. Clear is a color extender, and all these can be intermixed to create other colors. Keep all Kwik-Print colors tightly closed. Store them in a cool place.

KWIK-PRINT WIPES:

These sturdy cotton wipes must be used to coat the vinyl sheets or artist's paper. Cotton pads or Webril Wipes, however, can be substituted for those supplied by the manufacturer.

KWIK-PRINT AQUA AMMONIA OR HOUSEHOLD AMMONIA:

These solutions are used in the developing stage to remove unexposed color. Kwik-Print Aqua Ammonia (diluted 1 oz. per gallon or 8 ml per liter water) is a 14 percent solution of ammonia. Ordinary household ammonia can be substituted.

Caution: Ammonia is poisonous, and the fumes are extremely strong. It can be irritating to the skin, and medical attention should be sought if the solution gets in the eyes or is swallowed.

KWIK-PRINT BRIGHTENER

(Optional): Kwik-Print Brightener is stronger than the ammonia solutions because it is a reducer. If you are not careful, it will remove entire color layers. It is necessary to have this solution available when working with the Wide Tone base sheets because highlights are otherwise difficult to clear.

Kwik-Print is available in kits, or individual items can be ordered separately. The Kwik-Print Fine Arts Starter Kit contains six 5 × 9 base sheets (two each of Hi-Con V and Wide Tone V and P); three 4-oz. (118-ml) bottles of color (lemon yellow, medium blue, and magenta); one 4-oz. bottle of Aqua Ammonia; and one Kwik-Print coating block with thirty wipes. This kit is exactly what the name indicates; a starter sample.

The Kwik-Print Fine Arts Advanced Kit includes nine 8 × 10 base sheets, three of each kind; four 4-oz. (118-ml) bottles of lemon yellow, magenta, medium blue, and black; a 4-oz. bottle of Clear; one 4-oz. bottle of Aqua Ammonia; a 4-oz. bottle of Brightener plus the block and 100 wipes. This is perhaps the best of the kits to buy because the size of the base sheets is more workable than the 5 × 9 size, and the kit contains a bottle of Clear, which extends the colors, making them more transparent. The Brightener can be used for experiments with the Wide Tone sheets.

Two Color Master Kits are also available. One contains all fifteen colors and a bottle of Clear, all in 4-oz. (118-ml) bottles, and the other has all fifteen colors plus Clear in 8-oz. (237-ml) bottles. Neither kit contains base sheets, so they must be ordered separately. The Color Master Kits are an excellent buy if you plan to do a great deal of Kwik-Printing because they offer larger quantities of the colors at a greatly reduced price. Additional sheets, wipes, and other assorted products can be ordered to add to or replace items in the kits.

APPLYING KWIK-PRINT

The coating of Kwik-Print must be done in a dimly-lit room. A 50-watt yellow bug light about 6 to 8 ft. (1.8 to 2.4 m) from the coating area is sufficient for this step. The most successful results are obtained when the manufacturer's base

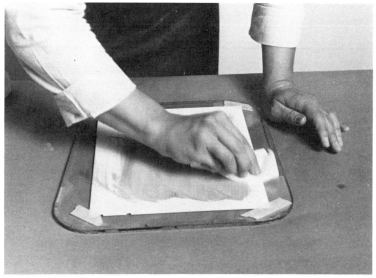

Secure a vinyl base sheet to glass, using tape if necessary, and squirt a small amount of Kwik-Print emulsion into the center of the sheet.

Quickly spread the emulsion, working it from edge to edge and corner to corner.

sheets are used. Varied results and problems arise when using artist's paper. Fabrics work well when certain specific steps are followed.

COATING KWIK-PRINT'S STABLE BASE SHEETS

The heavy-duty vinyl sheets are *not* light-sensitive and can be opened and examined in normal room light. The notches—coded *one* for Hi-Con V, *two* for Wide Tone V, and *three* for Wide Tone P—indicate the type and working surface. Handle these sheets like sheet film, placing the notched end so that it appears for coating in the upper right-hand corner. This way, the working side will face you at all times.

Secure the sheet to a hard surface—preferably a piece of glass. Tape the corners down or use a few drops of water on the back, pressing the vinyl sheet down hard until it sticks to the glass. This method allows you to use the complete im-

age area, working out to the edges of the sheet.

Next, fold a cotton wipe several times to make a pad, or wrap a wipe around the Kwik-Print block. Pour a small amount of Kwik-Print

color into the center of the vinyl sheet (see illustration). A puddle about 1 in. (25 mm) in diameter will cover an 8 × 10 sheet. *Quickly*, before streaking can occur, spread the color evenly over the sheet,

Buff the surface immediately after application, because Kwik-Print has a drying agent in it that will cause the liquid to harden almost instantly.

Within 15 minutes after coating, place the coated sheet in a contact frame with a negative and expose it to an ultraviolet light source.

first horizontally and then vertically. Immediately buff the surface with a clean wipe, spreading the excess off the edges. Continue buffing until the emulsion has dried. This may require the use of additional wipes.

Another method is to apply color directly to the wipe, then spread it on the sheet, first in one direction and then another. Quickly buff the vinyl surface before the emulsion has a chance to harden.

The coating should be even, with no streaks. If streaking does occur, immediately remove the coating with water, buff the surface dry, and coat it again with Kwik-Print color.

The manufacturer suggests using a color sequencing of dark to light. Registration of negatives for additional colors is easier if the first color is dark. Outlines of shapes are easier to see, and the colors are brighter in the finished print because the lightest color has been printed on the top. Do not use yellow last, because a mottling effect will occur, creating a distracting surface on the finished print, unless of course this is a desired result. (See Chapter 6, "Registration for Multiple Printing.")

COATING KWIK-PRINT ON ARTIST'S PAPER

Good artist's paper should be prepared by sizing before it is coated with Kwik-Print emulsion. Select a paper with some rag content. Rives BFK, Arches (both 100 percent rag), or Art Print (25 percent rag) will withstand long immersions in water. It is advisable to read the sizing section in Chapter 6 because the following preparation steps are the same.

Preshrink the paper in a bath of hot water. Hang the paper to dry. When it is completely dry, size it, using the unflavored gelatin/formaldehyde formula. Some photographers have gotten satisfactory results using the Elmer's glue mixture, but using unflavored gelatin guards against staining in the unexposed areas, and it will withstand forced development, other surface manipulations, and multiple-color printing.

Using a paddle brush or large paintbrush, coat the paper evenly, first horizontally and then vertically. *Quickly* buff the surface with a cotton wipe, spreading the excess off the edges of the paper. Hang the treated paper to dry in a dark place. Force-drying with a hair drier at the cool setting will speed

the drying. Remember, the paper must be completely dry before it is exposed.

COATING KWIK-PRINT ON FABRIC

Coating fabric with Kwik-Print requires some special preparations. First, unlike the other fabric emulsions presented in this book, Kwik-Print works best on *synthetic* fabric. Acetates, synthetic satins, nylons, and crepe readily accept the emulsion and will withstand the soaking in water. If you want to use natural-fiber fabrics such as cotton or linen, preshrink the fabric in hot water and size with either Elmer's glue diluted 1:3 or liquid starch diluted according to package instructions in a tray of warm water. Hang the sized fabric to dry, and iron out wrinkles before coating either fabric type.

Dilute the one part Kwik-Print color with two parts Kwik-Print Clear. If Kwik-Print color is applied straight from the bottle, the fabric becomes too saturated with emulsion and it is hard to clear the unexposed areas. You can add water to dilute the Kwik-Print color only if stiffness results after dilution with the Clear emulsion. Diluting with water can also cause staining. Consequently, adding water is a last resort for the control of this coating procedure.

Lay the fabric flat on a clean, hard surface. Using a large, coarse-bristle brush, apply the emulsion to the fabric. Make sure that it is absorbed into the fabric's fibers. Airbrushing is an alternative method

Forcefully squirt the print's surface with water.

of application. This avoids stiffening of the fabric, and there is less chance of staining in the highlights when this is done.

Hang the treated fabric to dry in a dark place. Using a blow drier will shorten the drying time.

EXPOSING KWIK-PRINT

Kwik-Print emulsion will fog if it is not exposed within 15 minutes after the coated surface has dried completely.

Select a negative that corresponds to the coated base sheet. Line negatives print best on Hi-Con V, and continuous-tone negatives work best on Wide Tone V or P. Either type of negative will work satisfactorily with artist's paper or fabric. Do not forget the alternatives to litho film as explained in Chapter 3.

Select an ultraviolet light source, such as the sun, sunlamp, carbon arc lamp, quartz, or photoflood. Exposures for Kwik-Print are very short compared with those for the other nonsilver processes. Each color has a different time because of its sensitivity to the visible light spectrum. The following table offers suggested times from the manufacturer for the three process colors and black, using a sunlamp 3 ft. (1 m) from the contact frame:

SUGGESTED TIME IN MINUTES

Black	Blue	Magenta	Yellow
2½	¾	1¼	1½

Note: Exposure on the Wide Tone sheets is critical, so a test strip is imperative before each color is printed. Hi-Con can be over-exposed and salvaged with Brightener with no apparent effect on the emulsion. Exposure times must be increased for artist's paper and fabric because of the greater saturation of emulsion on these surfaces.

Charting your own times for each individual color used with your specific light source will guarantee better results and allow more control of the process.

DEVELOPING KWIK-PRINT

This is a wash-out process similar to gum printing, but because of the stability of the Kwik-Print emulsion, the exposed print does not need to be handled so delicately.

Using a hose or water from the tap, squirt the Kwik-Print with a strong stream of water (see illustration). The unexposed color should wash off immediately. If color is still present in the highlights after 2 minutes, spray these areas with a weak mixture of household ammonia or the Aqua Ammonia (diluted 1 oz. per gallon or 8 ml per liter) using an inexpensive sprayer.

The surface of the Kwik-Print can be rubbed gently with one of the coating wipes to remove any excess color. Specific areas can be force-developed by rubbing the print with a finger, paintbrush, sponge, or even an eraser. Do not scratch the surface of the emulsion, especially the Kwik-Print on the base sheets, because you will be unable to recoat or reprint over a damaged area. If the chemical base has been changed, retouching with pencils or paints later is the only way to repair the print.

If washing and the use of ammoniated water do not clear the unexposed or highlight areas, Brightener must be used. Since this solution is a strong reducing agent, the Kwik-Print emulsion becomes vulnerable on its application. Careful handling of the print in this step is important. Do not rub or otherwise manipulate the print when using Brightener.

Flood the surface of the print with a thin layer of Brightener. Allow it to stand from 5 to 20 seconds, then gently hose off the surface with fresh water. Wash it for about 5 additional minutes. Blot the developed print dry, using a soft cloth, blotter paper, or newsprint, or hang it to dry. Do not force-dry the vinyl sheets with heat

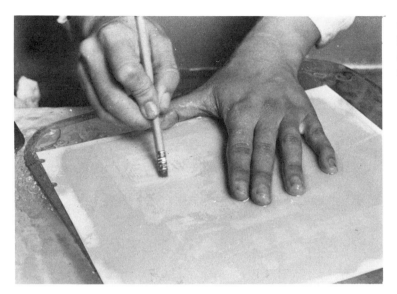

Forced development brings up more information in the highlights. Aggressive rubbing will not damage the image or the base sheet.

The Brightner is a reducing agent. Do not force develop the print when using the Brightner, or you may damage the image.

because this will affect the plastic base and could cause the emulsion to lift from the sheet.

The print must be free of water if you wish to apply a second coat of color. If you work in this expedient manner, the entire process of coating on the vinyl sheets, exposing, washing, blotting to dry, and re-coating should take only about 10 minutes per color.

SOME PROBLEMS—CAUSE AND EFFECT

Kwik-Printing is a simple process, but sometimes the results are not satisfactory. Being aware of the causes of these problems is important if you wish to master the technique.

An image that appears uneven after it has been developed is caused by sloppy coating of the emulsion. Areas of more thickly applied color are underexposed, and thin areas are overexposed. Apply the emulsion carefully, and be sure to buff the surface com-

pletely. If the color runs off easily and the details are lost during development, the print has been underexposed and exposure times should be increased to harden the emulsion.

If the image will not develop, even with the use of Brightener, either the print has been grossly overexposed, or the print was fogged because it was not exposed within 15 minutes after coating. Keep working the print with ammoniated water and/or Brightener. If all else fails, decrease the exposure time.

SEVERAL CREATIVE SUGGESTIONS

Because of the speed of coating and exposing Kwik-Print, full-color images can be generated quickly. Using four-color separation negatives you can layer color one on top of another to re-create a true color reality or fantasy. On Hi-Con vinyl sheets, over- or under-exposure alters an image greatly, so crude tonal separations can be constructed during the printing process.

Three or more coats may be applied, exposed, developed, and recoated. In the photograph above, a second coat is being applied over the first.

The title of Ken Steuck's Demonstration of the Art and Science of Physiognomy *is self-explanatory. Be sure to get your human subject's permission before you display or publish photographs in which the person is readily identifiable.*

The Clear emulsion extender allows for the softening of the colors, so subtle effects can be built gradually through multiple-color layers. The addition of other pigments, such as watercolors to the Clear extender creates subtle and even muted tonal renditions.

Selectively applying color to certain areas and printing them with the full negative will allow the building up of isolated areas of the image. Masking out areas of the print with goldenrod paper (an opaque graphic arts paper) is another method of working with the colors selectively.

Do not forget to try combining coating techniques using a paintbrush to build up textures with some layers, using wipes for others, and airbrushing for delicate areas. Applying color with pencils, inks, and paints, as well as creating a resist by working between color coats with wax pencils and/or crayons, will further enhance the image.

8

SPECIAL FABRIC PROCESSES

I became involved with nonsilver photography primarily because of printing on fabric. I first tried the paper-printing processes—blueprint, brownprint, gum print, and Kwik-Print—on fabric. Then I discovered special processes devised *only* for fabric printing.

There are three commercially made products, each with its own physical and aesthetic qualities, that are worth investigating. The most widely known is *Inkodyes*. These are transparent dyes which come in a full-color range of ultraviolet-light-sensitive solutions and are used primarily by fabric printers for the batik wax-resist process. Another product, Fabric Sensitizer FA-1, is a black-and-white fabric emulsion which contains silver and is manufactured by the Rockland Colloid Corp. (Fabric Sensitizer FA-1 replaces Rockland's Fast Enlargement Speed E-Mulsion BB-201.) The third specialty product is a presensitized black-and-white photographic linen called Luminos Photo-Linen. Rockland and Luminos both manufacture other specialty photographic products which are discussed in Chapter 9.

The manufacturers supply detailed instructions. All three products may be ordered by mail, although some large craft and camera stores carry a limited supply.

INKODYES

These permanent vat dyes are used by craftspeople, fabric printers, and artists both for the batik wax-resist process and for silk screening. The dyes penetrate the fibers of the fabric and are then set with either intense heat or exposure to ultraviolet light.

The dyes are suspended in a leuco base, which gives Inkodyes a syrupy consistency. They are transparent and can be intermixed. A clear extender is available to lighten the colors. Distilled water also dilutes the dyes' viscosity with only a slight effect on basic exposure times. When stored in opaque bottles away from heat or direct light, Inkodyes have an indefinite shelf life.

Single-color or multiple-color photographic printing on fabric can be accomplished with Ink-

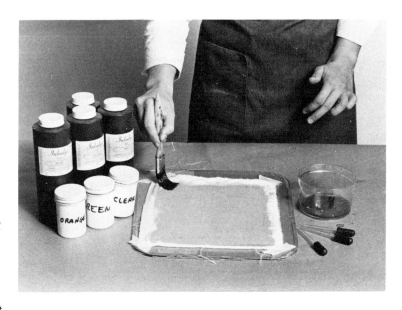

Secure the fabric to a clean hard surface. Apply Inkodye with a brush, working first in one direction and then in the other.

Place the fabric in contact with a large negative and expose to an ultraviolet light source.

odyes because they are controllable, consistently adhering to the fabric's fibers. If the dyes are diluted, subtle colors can be created to rival those of gum printing or Kwik-Printing. No sizing or special treatment of the fabric is required, but all printing *must* be done on natural materials such as cotton, muslin, silk, and artist's canvas. Polyester, nylon, and water-repellent fabrics reject the dyes, and so are not usable.

Exposure times vary, depending on the color. Development is completed in a tray of warm, soapy water, as will be explained later. After the fabric has dried, the dyes must be permanently set with heat. Inkodyes can also be printed on paper.

APPLICATION TO FABRIC

The following is the standard procedure for applying Inkodyes to fabric (see illustrations).

1. Secure the fabric on a hard, nonporous surface (glass or a Formica tabletop) with tape.
2. Under yellow safelight, pour a small amount of Inkodye into a glass or plastic bowl. The dyes will be similar in color (except yellow), so keep all bowls labeled to prevent mixing them when pouring the dyes back into their original bottles.
3. Use an inexpensive, large-bristle brush, so that the dyes will be forced between the weave of the fabric when coating. Staining will occur if too much pressure is applied; therefore, the paddle brush is not recommended. Do not apply pressure on the brush.
4. Evenly coat the fabric in only one direction, then reapply the Inkodye in the opposite direction. Hang the fabric to dry in total darkness.
5. When the fabric is completely dry, place it in contact with a large litho negative and/or positive, making sure both the surfaces touch. Expose it to an ultraviolet light source. Sunlight produces the brightest colors, but unfiltered ultraviolet bulbs, sunlamps, or carbon arcs also do well. Photofloods are too slow, and the resulting color values are not as pleasing. (*Note:* Wet areas expose unevenly and create uneven color. This is an

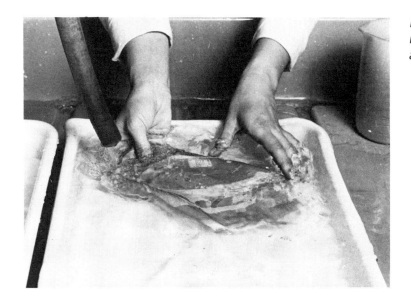

Develop in a tray of cold running water, vigorously rubbing and working the fabric.

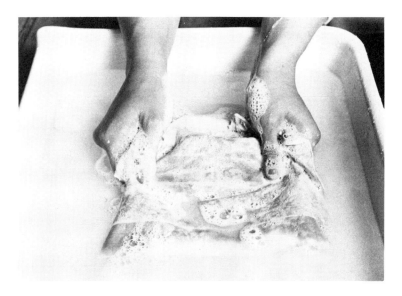

Continue development in a tray of warm soapy water, scrubbing out any stains in the highlights or unexposed areas.

interesting effect, but often the dyes harden onto the film, damaging it for any further printing uses. It is advisable to use fully dry fabric unless you are experimenting.)

6. An exposure test for each color is best for the light source in use, but the following approximate ratios among the undiluted colors (using a sunlamp 2 ft. or 61 cm away) will give you a starting point.

Yellow and violet are the fastest (approximately 1 to 3 minutes); red and blue take three times as long (approximately 5-10 minutes), and green and brown take six times as long (approximately 10 to 20 minutes).

Make a chart of the times using these ratios to save time later when printing multiple colors. Exposures should be increased slightly—about 10 percent—for diluted colors. Charting these and any intermixed colors will produce consistent results everytime.

7. Before development in ordinary roomlight, prepare two trays, one for cold running water and the other filled with warm, soapy water. (Suggestion: Dilute 1 tablespoon [15 ml] of liquid soap in 1000 cc tap water.) Develop by immersing the fabric in the tray of cold running water for 5 minutes. The dye from the unexposed areas will stain your hands, so rubber

Rinse in a tray of running water. Development and washing are now complete.

gloves are advisable. Change the water in the tray every minute, so that excess dyes run off and do not contaminate highlight or unexposed areas. Next, place the fabric in the tray of warm, soapy water, continuously agitating the tray for 5 minutes.

8. The fabric may be vigorously rubbed in the soapy water; this is sometimes necessary to clear the highlights from possible staining or overexposure. Some photographers actually develop large fabric pieces in the washing machine, first in a 10-minute agitation cycle without soap and then in a cycle with soap added. Complete with two rinse cycles.

9. At this point, if the dyes are still running, repeat step 8 until no more dye colors the water. Development is now complete.

10. Lay the image flat on blotter paper, and let it dry in this horizontal position. This allows the swollen dyes to set into the fabric's fibers.

11. When the fabric is completely dry, iron it at the cotton setting on the wrong side. Nontoxic fumes will result, and the color will deepen in value, but this final step is necessary to set the dyes permanently. It is unlikely, but dyes may offset onto the ironing surface, so it is advisable to protect it with brown paper or another piece of cloth.

To set the image, use a hand iron at the cotton setting. Apply pressure. The fumes are not toxic.

OTHER APPLICATION METHODS: SPRAYING AND DIPPING

The preceding steps are standard, but spraying or dipping the fabric can be substituted for steps 3 and 4.

The spraying method is preferable because highlights wash out easily, and slight errors of overexposure will not significantly affect the highlights. Colors remain subtle and more transparent than when the fabric is painted or dipped.

To spray Inkodyes on fabric, dilute them with warm distilled water until they reach a watery consistency. This will prevent clogging of the sprayer. Using an inexpensive sprayer available from a hardware store, spray the fabric first in one direction and then the other. Allow the first coat to become tacky before applying the second. Hang the fabric to dry. Exposure should be increased approximately 10 percent because of the dilution. All other steps are the same.

To dip fabric, fill a tray with dye diluted with distilled water. Again, a watery consistency is best. Wear rubber gloves. Immerse the fabric in the tray, and agitate it gently until the fabric is saturated. Drain off the excess, and hang the fabric in total darkness to dry. Experimentation is the best method of finding the results that suit your tastes.

HANDLING INSTRUCTIONS

Photographic images printed on fabric with Inkodyes are more durable than images printed by the blueprint, brownprint, or gumprint methods. Inkodye-printed fabrics can be incorporated into clothing, worn, and washed with no appreciable change in color. It is advisable, however, to wash them by hand in cold water, using Woolite or some other delicate detergent.

APPLICATION TO PAPER

Gum printing and Kwik-Printing are far superior to Inkodyes for full-color printing on paper. Although Inkodyes can be used, they are tricky; staining is a real problem.

Choose quality art paper with some rag fiber content, and size it (see Chapter 6). Use the spraying procedure discussed above, and apply the desired dye to a test strip plus a full sheet of paper. Dry them in total darkness.

Make a test strip and determine an exposure time. It is crucial to find the correct exposure time, because forced development will not remove staining. Expose the full sheet and develop it as though it were fabric, using cold running water in combination with a tray of warm, soapy water (steps 7, 8, 9, 10, 11).

Paper has a tendency to stain because the unexposed dyes drop out from the fibers of the fabric, but they do not release as easily from the tooth of paper. Rubbing or manipulating the paper's surface will not help. High-contrast film images render the best results. Trial and error are necessary when working with paper:

ROCKLAND'S FABRICS SENSITIZER FA-1

A second photographic emulsion, Rockland's Fabric Sensitizer FA-1, is commercially available to photographers and has many of the same characteristics as Inkodyes; however, this product is for printing only black-and-white *silver* images on cloth. It is a contact-speed emulsion requiring a large litho transparency for printing the image and a yellow safelight area for coating. Fabric Sensitizer must be applied to natural fabrics such as cotton, linen, and silk, but not to synthetics, because no colloidal binder is used. The latter is found in Rockland's other emulsion, *Liquid Light* (see Chapter 9). The advantage of Fabric Sensitizer over Liquid Light is Fabric Sensitizer's ability to lock into the fibers of the fabric.

Caution: Silver nitrate is poisonous. Keep the solution out of the reach of children. Wear rubber gloves and protective clothing to avoid absorption into the skin. Contact a physician immediately if accidentally swallowed.

Fabric Sensitizer is available in three packets labeled 1A, 1B, and 2. Packets 1A and 1B must be dissolved under dim incandescent light (a 25-watt household bulb 6 to 8 ft. [1.8 to 2.4 m] away) or yellow safelight, and only glass, plastic, or stainless-steel utensils are to be used.

Detailed instructions are supplied in the package. The following briefly summarizes the procedure.
1. Mix the contents of 1A with 1,000 cc of warm distilled or deionized water, and in another container mix 1B with 1,000 cc of water. Keep these solutions separate in brown bottles stored in total darkness until ready for use. The shelf life of 1A and 1B is only two days, and coated fabric cannot be stored. Plan to coat and print within this 48-hour time span.
2. Packet 2 is not light-sensitive and can be mixed in normal room light.

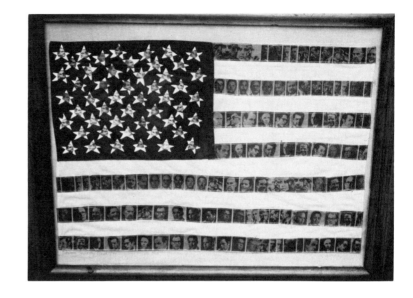

Linda Lindroth's Watergate Flag *is included in the collection of Paris's* Bibliothéque Nationale. *The flag is a mixed-media piece on a Photo Linen background.*

Dissolve it in 2,000 cc of cool tap water. (Distilled water is not necessary.)

3. In a large tray mix equal parts 1A and 1B. Dip the fabric and wring out the excess. Dry in total darkness.

4. Expose the fabric to an ultraviolet light source (sunlight, carbon arc lamp, or sunlamp) through a film transparency. This is a contact-printing process, so 100 percent contact is imperative for image clarity. The density of the film affects exposure, but 2 to 5 minutes is an average range to calculate first exposures. After exposure, the print appears medium tan.

5. Place the print in a tray of cool running water for 30 seconds to 1 minute. The image develops to an orange-brown color.

6. Fill a second tray with the Part 2 solution. Immerse the fabric for 5 to 15 seconds until the image turns gray brown.

7. Do not keep the print in Part 2 for an extended period or the image will begin to fade.

8. Wash the print in a tray of warm, soapy water (1 tablespoon of liquid soap to 1,000 cc of water) for 15 minutes. Rinse out the excess soap and lay the print flat to dry.

9. If you are not pleased with the image, it can be removed using a weak bleach bath (one part household bleach to five parts water); rinse thoroughly afterward. Then coat the fabric again, and repeat the procedure.

These fabric images can be toned blue, using any commercial photographic toner, without affecting the background. Other colors could be tried, but Rockland suggests only blue.

FABRIC SENSITIZER ON PAPER

Fabric Sensitizer FA-1 can be coated on paper as well for some unique results. Tonal gradations and subtle nuances of the emulsion's color on paper can be achieved by experimenting with the application of the solution. Size 100 percent rag paper first. This will ensure satisfactory results because the paper will then be able to withstand long periods of immersion in water. Airbrushing is the best method of applying Fabric Sensitizer to paper because the emulsion is evenly applied with little saturation into the paper's fibers. Applying the emulsion with a brush creates a painterly effect, however.

Allow the treated paper to dry thoroughly in total darkness. A hair drier, at a cool setting, speeds up this step. If the paper is not allowed to dry completely, some interesting things happen when the print is exposed to light.

The exposure and development times for paper are the same as for fabric. Washing time should be increased from 30 minutes to a full hour because the image fades later if it is not washed long enough. Fogging sometimes results, too, if washing is not sufficient. The soapy water should wash out any unexposed emulsion, but if chemical stains appear later, the wash time should be increased for the next print.

ADDING COLOR

Because Fabric Sensitizer is a black-and-white printing process, color can be added after the fact. Toning the paper with commercial photographic toners affects the silver image and the paper as well; staining may occur, but keep in mind that the results could be interesting. Applying watercolors to a wet but finished print creates a wet-on-wet effect or bleeding of the color. A metallic watercolor causes a bleaching-out effect when it is in contact with the emulsion's silver nitrate content. Hand

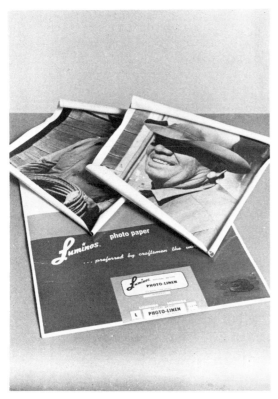

Photo Linen is precoated so that it saves you time and prevents the mess of coating it yourself.

coloring with pencils, chalks, inks, and various other methods of applied color are discussed in detail in Chapter 12.

Using Fabric Sensitizer requires some experimentation. Moreover, keeping records of results and solving problems for each individual image is essential. Although the accidents are exciting, real rewards come from learning to control the process.

PHOTO LINEN

Photo Linen is a commercially available product manufactured by Luminos. It is a continuous-tone, precoated linen that eliminates the need for coating tools, chemicals, and other items necessary for printing with Inkodyes and Fabric Sensitizer. When finished, Photo Linen handles like other photoprinted fabrics, accepting applied color, toning, sewing, and stuffing. It requires a minimum darkroom setup and ordinary black-and-

white chemistry. It is a projection bromide with grade 2 (normal contrast) emulsion. It is available in standard sizes such as 8 × 10 and 11 × 14 and can be ordered in larger sizes.

Although it handles like paper, it will not stand up to overdevelopment. Exposure of the image must therefore, be precise. Long periods in the solutions can break down the emulsion coating, so expediency in processing is recommended.

Photo Linen fogs easily from exposure to either light or intense heat. Refrigeration ensures longevity, but a cool, dry, dark place is sufficient to preserve it.

Use a test strip to determine the correct exposure time. Develop in Dektol, diluted 1:2, for 1½ to 2 minutes. Place the strip in a stop bath for 30 seconds, and fix it for 2 minutes before evaluating it in room light. Fix for a total of 10 minutes. Do not overfix; the emulsion could lift off or the image fade.

The fabric will be very limp during processing, and handling can be difficult. Placing plastic clothespins on all four unexposed corners will keep the linen from doubling over in the trays. Lifting the saturated fabric by the clothespins at the top corners when moving it from tray to tray will aid in ensuring even processing. (Wash the clothespins thoroughly before reusing them; otherwise, contamination will occur.)

Photo Linen should be washed completely, and a hypo-eliminating bath is suggested to cut this time. Without a hypo-eliminating bath, wash for 20 to 30 minutes.

Dry Photo Linen flat on aluminum screens or blotter paper. The fabric shrinks slightly as it dries, so tape or pin the borders down to prevent curling. If curling or extreme stiffening of the Photo Linen occurs, soak the image in a liquid fabric softener diluted one capful to a quart of water in a tray. This softens the curl and allows for easier handling after it dries again.

Photo Linen can be toned, hand-colored, stitched, and stuffed. The choices are unlimited for finishing the image. Additional possibilities are discussed in Chapter 12.

Make sure the image merits printing on fabric. An actual three-dimensional space can be created to enhance the photographic image already present on the flexible cloth. Consider stuffing and shaping, or machine stitching and hand embroidery to the image's elements when working. It is one thing to print an image on material; it is another thing to make it work successfully. Challenge your idea, visualize how it will look, and then get to work!

9

SPECIAL PROJECTION PROCESSES

Projecting images onto Liquid Light, Photo Aluminum, and other specialty products is a creative and experimental task that extends the boundaries of and challenges your ideas about photography. All these products contain silver. Each product's application, emulsion-base support, and end-results vary greatly from the others', making each a special and unusual artistic photographic process to consider.

ROCKLAND'S LIQUID LIGHT

Liquid Light is Rockland Colloid Corp.'s photographic emulsion for printing black-and-white images on any surface. Liquid Light can be coated onto chinaware, glass, acrylic, plastic, wood, rocks, eggs, artist's canvas, cloth, homemade and watercolor papers, jewelry, or any surface that is porous enough to accept it, and followed by normal black-and-white chemical processing. Liquid Light must be applied under amber, red, or yellow safelight. Its emulsion is sensitive only to the blue end of the visible light spectrum, so a dimly lit room will suffice. It requires an enlarger, trays, black-and-white chemistry, and other standard darkroom supplies.

The product must be stored in a cool, dark area, and for a longer shelf life, in a refrigerator. Do not freeze it because this will break down its viscosity. Unnecessary heat will "age" the emulsion, eventually causing it to fog. An antifog solution is supplied and can be added to correct this inherent characteristic. Fresh Liquid Light does not produce results as desirable in terms of density and contrast as an "aged" emulsion. Therefore, it is best to check the expiration date and to purchase, if possible, a bottle that has matured slightly. The emulsion can be forcibly "aged" by heating it to 140° F (60° C.) for an hour, and then letting it cool slowly before use. The test for sensitivity will dictate whether this step is required. (See "Testing the Emulsion," page 93.)

Warning: Do all coating and handling of Liquid Light in a well-ventilated room. The mixture contains phenol, and poisonous vapors can escape. Use rubber gloves and avoid prolonged contact with the mixture.

LIQUEFYING THE EMULSION

At room temperature, Liquid Light is solid. The emulsion must be heated to the liquid state before it can be applied to a surface. Immersing the entire bottle in hot water is the recommended procedure (see illustration). Set the bottle in water in a saucepan and heat until the bottle becomes too hot to touch. This should melt the emulsion without causing too much aging. Wait 3 to 5 minutes to be sure that a sufficient amount has liquefied, then pour off what you need into a glass, plastic, or stainless-steel bowl. Do not use copper, brass, bronze, or iron utensils, because the silver in Liquid Light reacts chemically with these metals. *Note:* Liquefy only enough emulsion for one session. The excess will harden in the bowl and must be discarded.

Remove the bottle from the hot water. If Liquid Light is left standing in hot water, it will age rapidly and cause fogging later.

APPLYING LIQUID LIGHT

Different surfaces require different methods of handling. Suggestions for coating paper, glass, metal, and fabric are provided to familiarize you with what to expect.

For coating paper, a paddle brush renders the best results with very little streaking or unevenness. Make this felt applicator as explained in Chapter 4, then dip it into the bowl until the felt is saturated.

Remember to select a paper that can withstand the photographic solutions and long periods of immersion in water. Rives BFK, Arches, and many of the other watercolor or printmaking papers work well with Liquid Light.

Apply the emulsion first in one direction. Be careful about setting the brush down too heavily at the start of each stroke to avoid puddling. Apply a second coat at right angles to the first. A gentle buffing of the surface with a soft cloth is recommended to smooth out the emulsion's surface.

If Liquid Light is too thick, it can be thinned by adding up to 20 percent warm water, but this increases the emulsion's drying time. Drying can be accelerated by a fan or hair drier set at a cool temperature. Do not force-dry with heat because fogging could occur.

These sensitized sheets cannot be stored overnight, so coat only enough to print and process in one session. Allow for test strips; two tests are usually required before the final print is exposed. Always work on a small scale before you make large images.

Painterly effects are created through the use of brushes. As with the processes discussed previously, the emulsion application should be considered in light of the content of the image selected. Brush strokes sometimes add to the photograph's meaning and give the image a new interpretation.

To coat glass, metal, or other small flat areas, a method of pouring and draining renders the best result. Quickly pour a large amount of Liquid Light onto the center of the coating area and tilt the object, allowing the solution to flow evenly over the surface (see illustration). Immediately pour off the excess before hardening can occur. Place a corner or edge of the surface back in the bowl, or use a funnel to drain the excess back into the bottle. Tap the coated surface

gently as you set it down *face up* to dry, so that any bubbles present will break. If this jarring action does not work, blow on the bubbles carefully so as not to disturb the surface.

To coat artist's canvas or heavy woven fabrics, a large, coarse-bristle brush works best because it forces the emulsion between the fibers of the fabric. Fabrics that have finer texturing should be dipped in a diluted solution and then hung to dry. (See Chapter 8 for Rockland's special fabric emulsion.)

DIPPING

For coating irregular or oddly shaped objects such as rocks, eggs, sculptures, or ceramics, the dipping method produces the best results. Dip, drain, and wait until the emulsion becomes tacky to touch, and then dip again and dry. Leave a small area uncoated, to simplify handling, and the emulsion will not be disturbed during processing. Remember to use tongs and/or rubber gloves while doing this.

Liquid Light can also be applied by spraying. Dilute the emulsion by mixing 100 cc of rubbing alcohol with 100 cc of water. Stir this combined solution into 1 pt. (473 cc) of Liquid Light immediately before application.

Caution: Phenol vapor is hazardous, and all spraying of Liquid Light must be done in a well-ventilated area. A respirator that absorbs organic vapors should be worn over the nose and mouth while spraying.

Small, refillable sprayer units such as Preval and Badger, which are available at most hardware stores, will work if small areas are to be sprayed. Larger areas require a compressor-type spray-gun unit equipped with external mix nozzles designed to prevent clogging. Check the nozzle because the silver halide in Liquid Light reacts with metallic surfaces other than stainless steel and causes contamination.

TESTING THE EMULSION

Two simple tests should be used before any serious work is attempted. One is a test for fogging; the other is for sensitivity.

Coat the surface of a sample of the material on which you wish to print the final image with Liquid Light. Without exposing this test strip to light, process it using the correct times. Fix it for a minimum of 5 minutes in a hardening fixer, then inspect it for gray or fogged areas. If the test strip does not appear completely clear, add twenty drops of antifog solution to 250 cc (8 oz.) of working developer. If this does not clear up the fogging, an additional twenty drops or more should rectify the problem.

To test for sensitivity, expose an image as you would a test strip for graduated intervals of time. Process. Evaluate to determine the correct exposure for the negative on the coated surface.

PROCESSING LIQUID LIGHT

Liquid Light emulsion is processed in a manner similar to that of a conventional black-and-white paper, except that all solutions must be at 70° F. (21° C.) or cooler because the emulsion will peel off in warmer conditions.

Develop the paper or object for about 1 to 1½ minutes in Dektol diluted 1:2 (remember to use an antifog solution if called for by your

previous testing). Obviously, three-dimensional objects cannot be tray-processed, so the developer must be applied with a soft sponge or atomized with a spray bottle. Proper development must then be determined by observation, not by a specific time.

Next, use either a commercial photographic stop bath or ordinary white vinegar diluted 1:2 for the same amount of time as the development.

Finally, use a hardening fixer for *double* the time it takes the image to clear, or a minimum of 10 minutes. As the image clears, the natural surface of the object will appear through the transparent areas.

Wash by using a hypo-eliminating bath for 2 minutes, followed by a total washing time of 10 minutes for nonporous materials (china or rocks, for example) and 30 minutes for porous surfaces (paper, wood, fabric). A constant flow of water must wash the surface of the object. Do not forcibly apply or squirt the surface with water because the emulsion will lift off. Keep the wash water at 70° F. (21° C.) or cooler.

Allow the paper or object to air-dry. While it is drying, handle the finished surface carefully because the emulsion is still vulnerable and could be damaged. As long as the hardened emulsion is kept dry, it is quite tough, but if it is exposed to moisture, it could lift from its base support. Coating the finished image with polyurethane varnish or some other protective coating should be considered if exposure to moisture is likely.

To clean the coating area and tools, simply use hot water. Stains can be removed from the hands with a 1:30 diluted solution of household bleach in water, fol-

lowed by a rinse of fixer and then a wash with hot, soapy water. Detailed instructions are supplied with Liquid Light, and troubleshooting hints for various surfaces and printing problems are also provided.

OTHER PRODUCTS

Rockland Colloid Corp. also manufactures Photo Aluminum, precoated, light-sensitive aluminum sheets available in standard sizes. They can be cut to meet specific enlarging needs, and they are processed in conventional black-and-white chemistry similar to photographic papers. Using Photo Aluminum is the easiest procedure for putting photographic images on a metallic surface. The surface quality of the metal, its reflectance, and the possible working of the metal itself (embossing and engraving) will further extend the meaning of the image.

Rockland revived an old process with the Tintype Kit. This outfit contains five 4 × 5 presensitized sheets and the necessary chemistry to produce vintagelike tintypes.

Rockland also supplies two toners for this specialty market: PrinTint and Halo-Chrome. PrinTint tones or colors only the white areas of a black-and-white silver print and does not affect the silver image. Halo-Chrome changes the silver halide in a black-and-white print to metallic silver, which will tarnish without sufficient protection. Both toners are discussed at length in Chapter 12.

LUMINOS PHOTO PAPERS

In addition to Photo Linen (as dis-

Pasha Turley's Elnora *is a photographic print on canvas that has been sensitized with Liquid Light. Turley finished the image by hand coloring local areas.*

cussed in Chapter 8), Luminos manufactures a line of specialty photographic papers in various pastel colors, such as green, blue, yellow, pink, and red. These single-weight papers are available in 8 × 10 size, in normal no. 2 contrast grade, and are processed by regular black-and-white chemistry.

There are some minor problems with these papers. Under safelight, it is difficult to determine the emulsion side of the paper. The best way is to set the paper down on the baseboard of the enlarger and determine which side curls inward. This indicates the emulsion side.

Another problem arises when determining the proper development time. Dektol diluted 1:2 is the recommended solution, and 1½ to 2 minutes is a standard minimum development time. Because of this product's pastel paper support, the image appears to develop quite quickly, and there is a tendency to stop development before it is complete. Stick with a predetermined time because in room light the print tends to appear completely different. Consequently, valid evaluations must be made under white lights—not in the darkroom.

Another problem is that the print may appear harsh. This is caused by the black and gray image on the light pastel paper. Sepia toning warms the image and renders the print more pleasing. Other color toners can also be used with these papers and should be considered for creating different effects. Many color relationships can be explored to produce exciting results. Moreover, many of the darkroom manipulations used with conventional black-and-white papers can also be tried with these papers. Double printing and solarizing are just two possibilities.

Whatever you choose to do, be sure that the negative selected will work well with the pastel color of the paper. Keep in mind that no special process will improve a poor negative choice.

BLENDING PROCESS AND IMAGE

Many photographers, including those who use nonsilver processes, consider some, or all, of the materials discussed in this chapter as gimmicks. They can be if the image and how well it relates to the process selected is not considered. For example, printing flat, nondescript images on rocks with Liquid Light is self-defeating. Instead, choose a negative that enhances the combination of a photographic image with the shape of the rock. These may seem like obvious and simple considerations, but they are important ones if you wish to push these processes beyond the gimmick level and into more serious photographic image-making.

10

MAGAZINE TRANSFERS and LIFTS

Throughout this book, unconventional processes and methods, by the usual definitions of photography, have been explored. Contemporary photographers working with these alternative processes have no strict definition of what their photographs are or what a photograph should look like. Many refer to themselves as "visual guerrillas"—taking, using, and reusing photographs and photographically produced images to create new visual and personal statements.

During the early 1960s the influences of Pop Art, the work of Andy Warhol, and the lithographs of Robert Rauschenberg were major factors in this new examination of the "photograph." Christopher Finch in *Pop Art: Object and Image* (E.P. Dutton and Co., 1968) writes, "An artist who uses a photographic image is not merely avoiding the issue of having to reproduce a subject manually. He is introducing onto the picture surface something which has, for the mass media oriented audience, an objective existence."

Along with blueprinting, brownprinting, and gum printing, photographers such as Darryl Cur-

ran, Bea Nettles, and Robert Fitcher began to explore the possibilities of reusing and incorporating already existing photographically derived images. Instead of color separating and reprinting the four colors, Curran chose to transfer the printed inks directly from magazines to the surface of his blueprint. This is done by a simple process called magazine transfer or rubbings.

MAGAZINE TRANSFERS

Almost any image from a printed source can be transferred to paper. The best source of material for transfers is a full-color Sunday newspaper supplement. The inks are highly saturated and freshly printed. The paper stock is a lightweight newsprint, and if you wish to transfer duplicate images, you can purchase cheaply several newspapers. The second-best choices are inexpensive magazines such as *Family Circle* and *Good Housekeeping*, because they have the same basic paper and ink characteristics as the Sunday newspaper supplement. *Newsweek* and *Time*

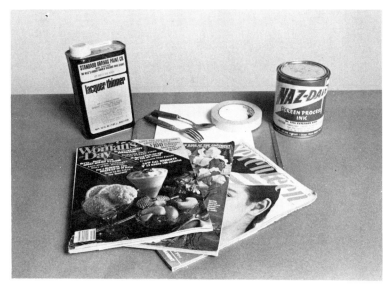

The supplies shown here are needed for magazine transfers.

will work, depending on the transferring solution used. Magazines such as *Playboy* or *Cosmopolitan*, that are printed and then coated with a layer of varnish, will transfer, but with a great deal of difficulty. These magazines require some experimentation and patience, but satisfactory results are possible.

The intent of this process is to transfer the printed inks from their magazine-page support to another receptor, the paper. To accomplish this, a solution must be applied that will dissolve or move the inks chemically from one surface to another without destroying the image.

TRANSFER SOLUTIONS

Silk-screen extender base, such as Naz-Dar, no. 5536, is similar in consistency to a syrup or a heavy oil. It can be spooned out of its container easily and applied more selectively than other transferring solutions. Lacquer thinner, lighter fluid, and Carbona Spot Remover are also good transferring solutions, but the combination of their fumes and their watery consistency make them less desirable when working for long periods of time. Lacquer thinner is best for transferring from the coated stock of such maga-

zines as *Playboy, Cosmopolitan, Time,* and *Newsweek;* silk-screen extender base works best with uncoated papers.

Warning: Good ventilation is necessary for working with all transfer solutions, but especially with lighter fluid, lacquer thinner, and spot remover. Read all directions on the container. Do not use transfer solutions near heat, fire, or open flame.

Select a smooth-surfaced paper as the receptor for your transfer. The paper should be durable. It must accept the pressure applied by the transferring tools. Moreover, it should be porous so that transfer of the inks is possible. A heavy card stock, a 100 percent rag fiber printmaking paper, the back of double-weight fiber photo paper, or even heavy-duty brown paper bags will all work.

The best transferring tool is an ordinary metal teaspoon. A printmaking burnisher used for metal etching plates is also a good tool with which to experiment. Metal forks, the flat handle of a knife, or any other smooth, rounded utensil is also a good choice. Search through your kitchen, garage, or local hardware store for interesting devices. A clean, smooth working surface is also necessary for proper

contact during transferring. A piece of glass on a sturdy tabletop is best, but a Formica countertop will do.

TRANSFERS USING SILK-SCREEN EXTENDER BASE

The following steps explain how to transfer a magazine image using silk-screen extender base:

1. Tape the paper onto a smooth surface with masking or drafting tape.

2. Securely tape the magazine image you wish to transfer *face down* on the paper.

3. Scoop out with a teaspoon and place a generous amount of silk-screen extender on the back of a magazine page (see illustration). Spread it out over the entire magazine surface *only* and allow to stand for about 30 seconds to 1 minute. (This short waiting period allows the extender base to work chemically *through* the top surface of the inks as well as through the paper fiber of the magazine page.)

4. Rub the surface, working vigorously, either in a random motion or in a repetitive pattern. The inks on the top surface of the magazine page should begin to dissolve, and the extender base will become saturated with these dissolved inks and varnishes. Periodically, lift one end of the magazine page to see the degree of transfer.

5. Continue step 4 until the desired results are achieved. The pattern of

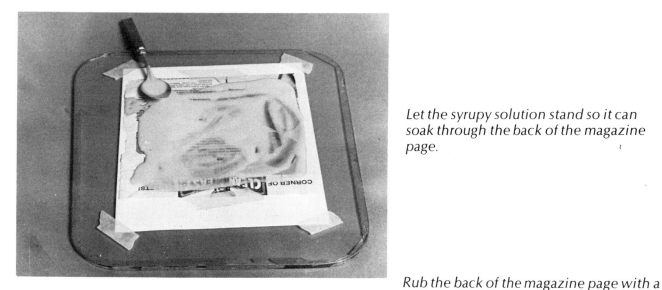

Let the syrupy solution stand so it can soak through the back of the magazine page.

Rub the back of the magazine page with a hard utensil.

the rubbing strokes can enhance an image as well as contribute to the intensity of the transfer. *Note:* If you wish a rich, intense image, allow the extender base to soak longer in step 3; up to 5 minutes is advisable.

6. When the transfer is complete, wipe off the excess extender base and discard it. Untape and remove the magazine image, and throw it away. Evaluate the image.

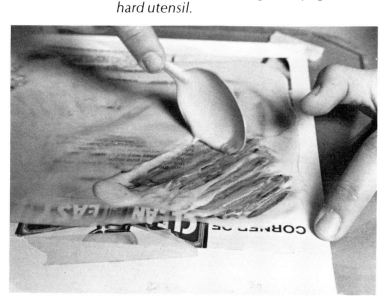

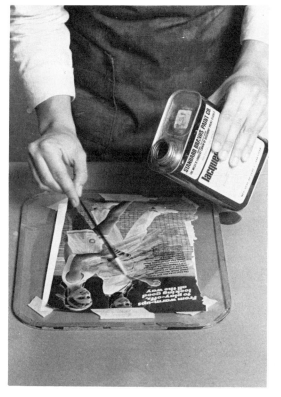

Use a small paint-brush to apply lac-quer thinner.

TRANSFERS USING LIQUID SOLVENTS

Work quickly when transferring magazine images with lighter fluid, lacquer thinner, or spot remover, because these solutions evaporate within minutes.

1. Repeat steps 1 and 2 for the silk-screen extender base process.

2. Pour or squirt *one* of these solutions, thoroughly saturating the back of the magazine page (see illustration at left).

Apply even pressure and work fast because the lacquer thinner will evaporate quickly.

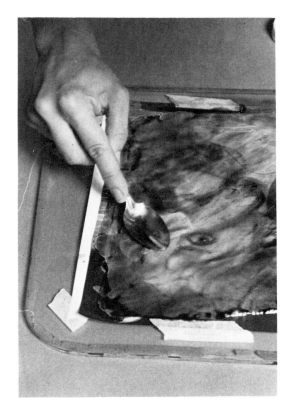

Re-wet and rework until desired results are achieved.

3. Immediately begin rubbing, applying pressure, with a metal teaspoon or other utensil. (*Note*: It may be necessary during the procedure to wet the back of the magazine page again with fresh solution.)

4. Be careful when lifting one edge of the magazine page to check the degree of transfer, because these solutions can dissolve the adhesive of the masking tape, causing the transfer to slip out of register.

As with the silk-screen extender base, staining will occur if the dissolved ink mixture is allowed to overflow onto the paper. You can prevent this by masking around the edges with bond paper or some other lightweight absorbent paper. When the desired effect is reached, remove the magazine page and throw it away. Seal the transfer solution tightly to avoid loss by evap-

oration, and store it away from heat and flame, or as instructed by the package directions.

OTHER TRANSFER METHODS

An alternate method for transferring which eliminates the characteristic rubbing marks requires access to a lithograph or etching press and the use of lacquer thinner. If either of these is available to you, set the plate-printing pressure to maximum or as advised by a printmaking authority. Place the paper on the bed of the press, supported by either a smooth, unmarked litho stone or a metal plate, depending on the press. In a metal tray of lacquer thinner, soak the magazine page thoroughly for

about 60 seconds. Drain it, and immediately lay the image face down on the paper. Cover it with a protective sheet of bond paper and the press blanket. Run this package through the press, applying maximum pressure as quickly as possible. (Remember that lacquer thinner evaporates instantly.) Remove the blanket, cover sheet, and magazine page. The image should have transferred evenly. Printmakers have used this same process to transfer magazine images to lithography limestones.

I have successfully used both the rubbing and the press methods to transfer images. After the transfer is complete, normal printmaking procedures are followed to prepare the stone for edition printing. Transfers expedite the printmaking process and allow for almost immediate image-making.

Apply the first coat of gloss polymer evenly, covering the entire image you wish to lift. Apply a second layer in the opposite direction and repeat the procedure until 10 coats have built up.

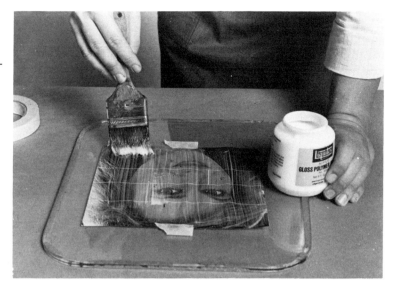

Immerse the coated image in a tray of hot water and slowly peel off the back of the paper.

CREATIVE ASPECTS

Evaluation of magazine transfers is, of course, a subjective matter. The direction of the rubbing marks, the intensity of the transferred inks, the reversed image, and the relationship of the image to the paper surface—all these aspects must be considered. The freedom of the process is overwhelming. Without a darkroom or camera, and with just a few supplies, you can generate photographic statements.

MAGAZINE LIFTS

In contrast to transferring, in which the printed photographic image is "rubbed" into the fiber of the paper, this process lifts the ink off the magazine page into a flexible plastic support. The selection of magazine sources, as described earlier in this chapter, applies here as well; inexpensive publications render the best lifts.

There are two basic ways to lift an image. One uses a painter's solution found in art supply stores, and the other uses clear Con-Tact paper. Both are quite successful methods; determining which to use depends on the desired result.

LIFT METHOD No. 1

Purchase gloss polymer medium, a milky-white acrylic paint extender that dries clear, and a 1- or 2-in. (25- or 51mm) inexpensive paintbrush. You need a nonporous, hard surface on which to work, because this solution is difficult to remove and clean up. A piece of glass is best.

1. Trim the magazine image, removing unwanted areas by cutting or tearing, and place it face up on a glass surface. Do not secure it with tape.

2. Dip the brush in the gloss polymer medium and apply the first coat evenly in one direction (see illustration).

3. Allow this application to dry thoroughly.

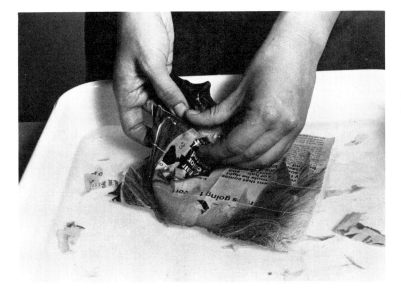

After soaking, remove the paper backing.

The gloss polymer lift is quite flexible and can be stretched and distorted.

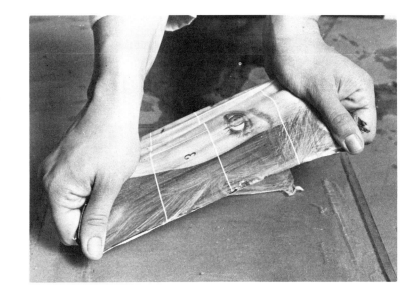

4. Repeat steps 2 and 3, brushing in opposite directions for each separate layer, until the magazine image has received ten coats.

5. After the tenth application has completely dried, fill a photo tray with hot water. Boiling water works faster, but hot tap water is sufficient.

6. Immerse the coated magazine image in the tray, face up. It reacts to hot water, turning a milky white, but it will clear.

7. Allow the image to soak for about 1 to 2 minutes. As the water cools, turn the coated magazine lift

over and carefully peel the paper fiber away if it has not already dropped off.

8. When all paper fiber has been removed, carefully lift the plastic image from the tray and place it on a clean piece of glass. Shape it if necessary and allow it to dry.

The inks from the magazine page are now securely embedded in this plastic sheet, creating a semitransparent photographic image. Lifts from gloss polymer medium are flexible and can be distorted by pulling and stretching.

LIFT METHOD No. 2

Use *clear* Con-Tact or other clear vinyl. Measure carefully and buy enough to cover all the magazine images you wish to lift. Then proceed as follows (see illustrations).

1. Repeat step 1 of method 1.

2. Cut a piece of clear vinyl a little larger than the magazine image. Carefully start to peel back the paper backing, holding the vinyl's top edges apart. (This is a difficult step, because this paper sticks to itself.)

3. Place the edges on the glass, over the magazine page, and press

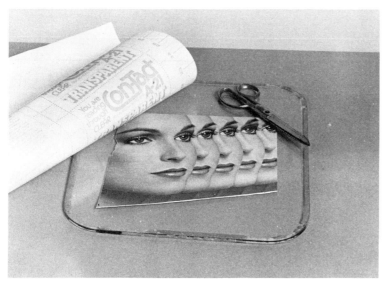

Clear Con-Tact paper, a magazine image, and a pair of scissors are all you need for a magazine lift.

Peel Con-Tact paper away from its support.

Adhere the top of the Con-Tact paper to the glass, and slowly apply pressure as you lay the adhesive side on top of the magazine page.

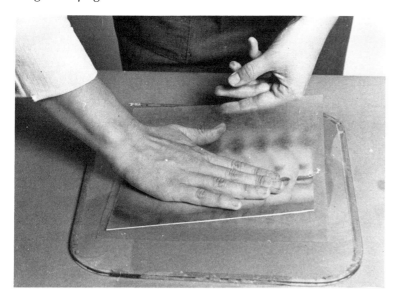

down so that the vinyl sticks to the glass.

4. Using one hand, slowly peel off the rest of the paper backing as you use the other hand to flatten the sticky vinyl over the magazine page.

5. Now vigorously rub the top surface with a metal spoon, etching burnisher, or a hard rubber brayer. Work in one direction, then the other, covering the entire surface. You can use a small needle to pop any air bubbles caused by uneven contact. The inks are now embedded in the adhesive layer on the vinyl.

Using a spoon or other hard utensil, apply pressure over the full surface.

Immerse the Con-Tact paper and image in a tray of hot water and allow them to soak.

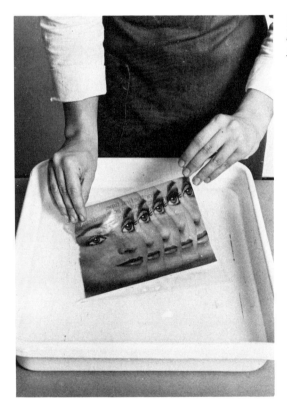

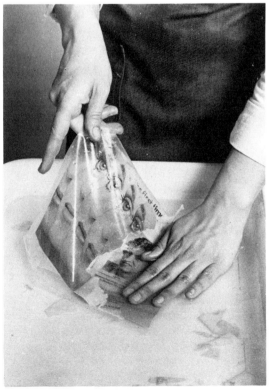

Gently peel off the paper backing.

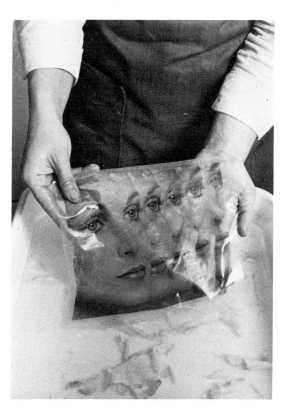

Remove the lift and blot it dry.

6. Fill a chemical tray with hot water. The same recommendations apply in step 5 of method 1 for this procedure.

7. Remove the vinyl and magazine page from the glass. Immerse the sheet in the tray, face up. Allow a 1- to 2-minute soak, then carefully turn the vinyl over and peel off the paper fiber.

8. Remove the vinyl and hang it to dry.

Vinyl paper lifts can be pulled slightly out of shape, but they are not as flexible as gloss polymer medium lifts. (*Note:* Save the white paper backing because the lifts can be placed on them, rolled, and stored easily.)

DECALON

A decal medium called Decalon is also available. It is marketed and distributed primarily through hobby and craft stores for making decals, but many photographers use it for magazine lifts.

Decalon is easy to use, and the instructions are very similar to those for Con-Tact or vinyl shelf paper. It has one advantage over shelf paper in that it can be stretched to twice its size without splitting or cracking. Distorted images can be created similar to those made by the gloss polymer medium method.

APPLICATIONS FOR MAGAZINE LIFTS

The applications for magazine lifts are many and varied. Both methods described generate an image that is semitransparent and can be used as an overlay with magazine transfers, conventional silver prints, or nonsilver-generated photographs.

As an overlay, the Con-Tact paper lift has a "slicker", cleaner presentation, but the greater distortion capabilities of the gloss polymer medium lift should not be overlooked for creative exploration.

Instead of imaged litho film, magazine lifts can be used directly as transparencies for blueprinting, brownprinting, gum printing, and Kwik-Printing. The result from these lifts will print out in the negative with very little contrast. To achieve a negative from a magazine lift, as well as increase its contrast, lifts can be contact-printed with litho film. (This process is discussed in Chapter 3.) Gloss polymer medium lifts can be stretched and reshaped to alter the image further prior to printing in any of the processes described previously.

Magazine lifts allow for working through ideas quickly, tapping a large number of visual sources without involvement in the picture-taking process, but you must consider copyright laws when using these and other copyrighted images.

The magazine transfer shown here was accomplished by using a lithography press. Suda House, the creator of the image, transferred the magazine's inks to a lithography stone, which was then etched and re-inked. The work is printed on Rives BFK 100 percent rag paper.

Todd Trigiani created this image by combining transferred magazine images on artist's paper.

Bobbie Kitchens's study of a nude began with a print on artist's paper sensitized with Rockland Colloid Corp.'s Fabric Sensitizer FA-1. She then hand colored the print to produce the delicate hues apparent in the reproduction above.

Danniele B. Hayes constructed this large (48 × 72 in.) hanging out of individually printed panels. Each section was generated by using Kwik-Print on nylon.

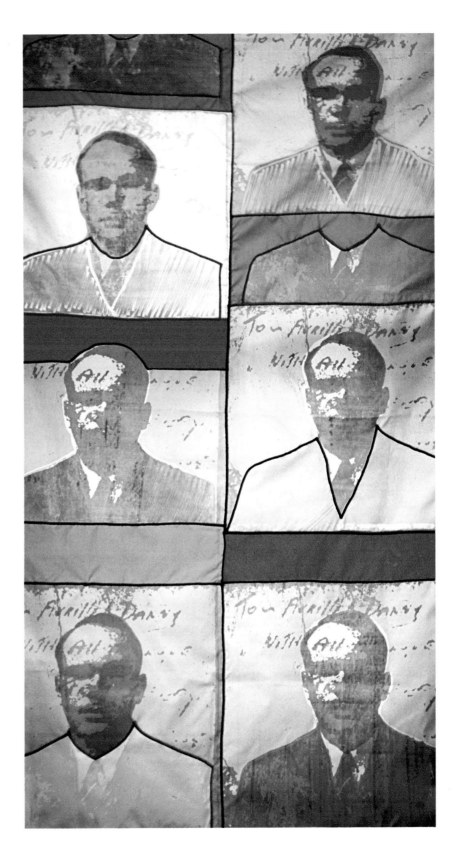

Jack Butler hand colored a Polaroid SX-70 print with Marshall's Photo Pencils to produce
Sex 70s. SX-70 prints, resin-
coated silver prints, and other
glossy finishes will accept hand-
coloring media if the surfaces of
the prints are first sprayed with
Marshall's Pre-Color Spray.

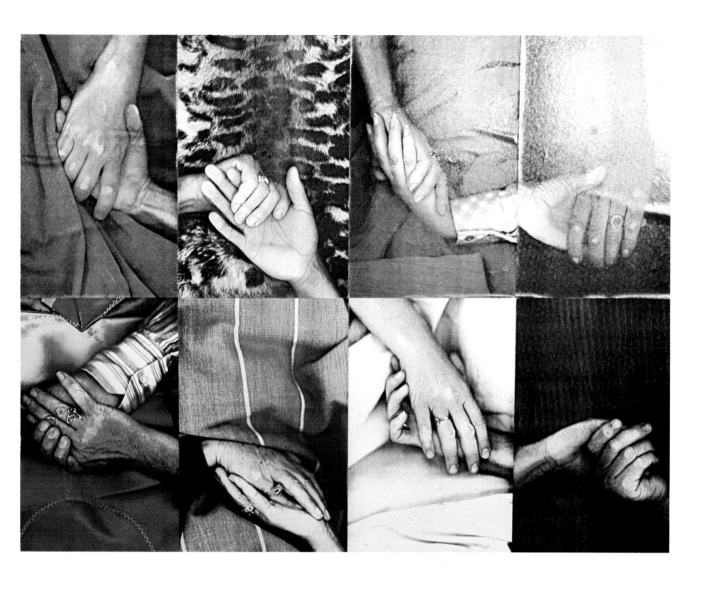

Pasha Turley asked both friends and strangers to hold hands with her while she made a Xerox color copy. After a month of experimentation, she put together her documentary piece Just a Hand to Hold.

California Alphabet, *by Jane Zingale, is a 42 × 44-in. piece created on a color Xerox machine. The original work was presented in two rows of thirteen pictures each.*

Marka Hitt used Xerox's special transfer process to accomplish Landscape #1. After transferring the image to paper with a hot iron, she reworked the Xerox print with pastels and acrylic paints.

Bobbie Kitchens used oil pastels and colored pens and pencils on a silver print to create Fantasy. She first sprayed the print with Marshall's Prepared Medium Solution so that the applied color would be absorbed into the print's matte surface emulsion.

11 | Copy Machines

INSTANT IMAGES

"Technology and art need not be strangers, nor at odds with one another."

This is the first line of the catalog for "Electroworks," a traveling exhibition of the International Museum of Photography at George Eastman House, which showcased works created on electrostatic copy machines. Many contemporary photographic image makers consider working with these machines and their results to be a nonsilver photographic process. More than a dozen years ago, when numerous photographers throughout the country began doing serious work with copy machines, many critics laughed at what seemed to be an instant-image. Now, however, museums such as the George Eastman House, are organizing exhibitions based on the short history of man's art-making with the office copier. Works in this show were created with machines such as 3M's VQC and Thermo-Fax and its now discontinued Color-In-Color copier; Xerox's numerous black-and-white copiers, plus their early Haloid camera; and the Xerox 6500 Color Copier, now in widespread use.

EDUCATED PLAY

"The machine is endowing art with an element of play—play in the sense of scope and freedom, play in the sense of sheer fun," said Douglas Davis in *Art in America* (Norton, 1968, p.46). Copy machines are fun. They are precise, instantaneous, and automatic. The first time I worked with the 3M Color-In-Color copier I had the same magical feeling as I did when I watched my first silver print develop.

Copiers provide immediate feedback of whatever you choose to reproduce. Three-dimensional objects may be set on the copier's glass, and some machines can make prints of 35mm color transparencies.

Whatever the imagination conjures up can usually be created. From the simple one-on-one duplication of a print or a slide to the transfer of these copies to papers or fabrics, the possibilities are endless.

Unlike the other processes described earlier, where the measuring of chemicals, applying of emulsions, and registration of negatives demanded mastering certain technical skills, copy machines require only one basic element—educated play. Technical precision is inherent in these machines; you supply the human element of *creativity*.

BLACK-AND-WHITE COPIERS

The place to start is the black-and-white office copier, which can be found everywhere from post offices to print shops. These machines are simply large-format cameras with a lens, built-in light source, and instant printing system, usually using electrostatic toners. In photographic terms, critical focus of the lens is on the glass or platen, where flat copy is placed for duplication. Many of these machines have a depth of field that extends up to 1 in. (24mm). Contrast (light to dark) can be controlled on some machines like the 3M VQC, although others, like the later Xerox models, have only a "Light Original" button that adds some density. A few models have reduction and enlargement capabilities.

The paper used with these different systems varies, depending on the manufacturer. The 3M copiers require the use of special papers coated to accept 3M's special toners. On the other hand, Xerox models can accept lightweight bond papers up to the thickness of thin card stock. Although typical copy size is 8½ × 11, some machines can generate copies as large as 8½ × 14 and even larger.

Almost all of the black-and-white machines can make copies on acetate sheets. These are often used for overhead viewer presentations, but artists have found other applications for them.

On some machines like the 3M VQC, the light source scans directly underneath the platen; the Xerox 914 has a slit-scan shutter that works in a similar manner. Both systems allow for movement distortion similar to that of the camera technique of panning. All these characteristics should be considered before you start exploring the copier as a serious photographic process.

THE CREATIVE APPLICATIONS

A copier vocabulary is slowly emerging among photographers and artists. Applications and ideas discussed here, using that vocabulary, have developed over the past decade; they provide a place from which to begin exploring the creative use of copy machines.

GENERATION LOSS

Generation loss refers to the result of making a copy of a copy of a copy. This produces a tone and detail loss from print to print or "generation to generation." The slow destruction of resolution can completely alter an image and eventually redefine its meaning.

Another approach derived from generation loss is the ability of some black-and-white machines to create crude tonal separations. On the 3M VQC a black-and-white print's varying contrasts can be extracted by changing the contrast-control dial from light to dark. Each turn of the dial generates a new printout of slightly different tonal saturations. These "separations"

can be contact printed onto litho film and used with such full-color processes as gum printing and Kwik-Printing.

Some photographers copy these "crude" tonal separations onto acetate sheets, then contact-print the acetate positives *directly* onto litho film to make tonal separation negatives. The powdered toners also create a random dot pattern that causes some interesting results.

OVERLAY

The acetate copy itself is yet another creative application, both as an overlay to another image—a blueprint or brownprint, for example—or by itself. Keep in mind that direct contact printing of an acetate copy onto sensitized surfaces such as cyanotype can create unique images, too.

Because these images are on clear plastic, they can be painted on the back, scratched, or otherwise manipulated. Animators have used this exact procedure of making copies on acetate for many years to expedite the animation process. Hand-colored, reworked acetate "cells" are registered and photographed by an animation camera to make animated cartoons.

SCANNING

Scanning with the light source or lens as it pans the ground glass creates blurred images. In general, mysterious, optically distorted images are created. Moving three-dimensional objects alters their appearance and changes their meaning as well.

TRANSFERS TO PAPER

Transferring any of these copies to another paper surface by the use of lacquer thinner is another creative form for this process. Because the size of the copy, that is, 8½ × 11 to 8½ × 14 is limiting in the sense that all copies leaving the machine must conform to the prescribed sizes, transferring to larger paper, collaging, and integrating various images make new statements that allow you to break from the machine format and size.

TRANSFERS TO FABRIC

Transferring to fabric, sewing, and stuffing the results into sculptural or relief images are further ways of utilizing the black-and-white copier's creative aspects. Practice is needed in attempting to transfer these copies to fabric, but some good results can be achieved.

Use a hand iron at its highest setting, apply even, hard pressure to the back of the copy placed face down on the fabric, then remove the paper quickly before it sticks. This usually produces a sufficient transfer in which intense heat releases the fused toner and the pressure forces it into the fibers of the fabric. A dry-mount press at 325° F (163° C.) can be tried, but unless its pressure or tension is greatly increased, the results are not always pleasing.

Black-and-white copying machines, because of their accessibility and low cost, offer the oppor-tunity to investigate the creative aspects of copier art. But trying one's imagination out on a color copier becomes yet another challenge and an adventure as well, because these copiers supply the photographer with full color.

COLOR COPIERS

"The creative age of color is coming alive. . . ." This headline, in a 1968 3M brochure, heralded the age of color copying. That year the 3M Co. introduced the Color-In-Color system that used two machines, one capable of reproducing flat artwork at a 1:1 ratio and the other a color-slide machine capable of copying positive transparencies from 8mm to 2 in. (50mm) square, and enlarging them to five, seven or eleven times their original size.

Both machines utilized electrostatic and thermal dye transfer systems to generate a dry, full-color print in 30 seconds. Artists soon discovered the striking color of the 3M machine. In fact the machine at times seemed to produce color that was more intense and more pleasing than that of the original.

These seemingly magical machines were pulled off the market in 1976 because of marketing and productivity problems. Color-In-Color was phased out, and is no longer available for public use. Some machines are available at private companies, but all copy centers are closed. Their extinction was hurried by the invention and introduction in 1973 of the Xerox 6500 Color Copier.

COLOR BY XEROX

The Xerox Color Copier has its own special characteristics and applications, making it an excellent full-color instant-image process for today's photographers. Xerox manufactured its product primarily to meet basic printing, business, and marketing needs, but what becomes interesting is how the artist

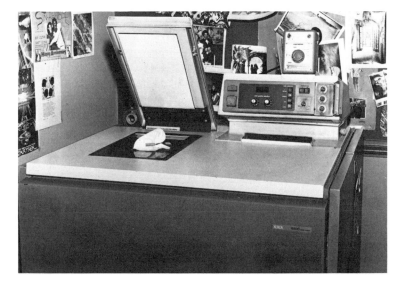

The Xerox 6500 Color Copier is very similar in appearance to an ordinary black-and-white machine. The difference is in the control panel at the right.

The addition of the slide projector and the screen modifies the 6500 copier for duplication of slides and other transparent color originals.

The three color dials are at the extreme right and the contrast control dial is to their left. The color deletion buttons are in the center of this panel.

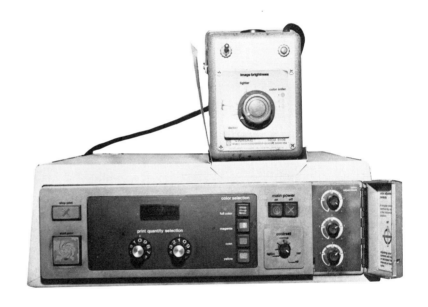

and the photographer adapted the machines to their own creative needs.

The Xerox color machine can copy flat art at 1:1, up to 8½ × 14. With the slide projector attachment and a plastic dot screen placed on top of the platen 35mm slides can be enlarged to approximately 7 × 9. This copier generates a clean, hard-edge copy and duplicates detail as small as characters on a typewritten page. The early machines produced copies with high-contrast color, but this was improved with the addition of a color-contrast-control dial. The copy itself is textured because of the dusting and layering of the paper's surface with toners—one each of the three process colors (yellow, magenta, cyan). These toners are then fused by heat.

Full color is not the only option

of the Xerox Color Copier. An original can be copied in a single color of the three process colors. Combining yellow and magenta produces an orange red, and lowering the contrast-control dial softens the red to a lighter shade. Yellow plus cyan make green, and magenta plus cyan make a bluish purple.

The color-contrast control does lower the overall density, but the saturation of each process color can be controlled by three separate color-adjustment dials. Each dial has a saturation range from 1 to 5.

Three is an average saturation selection, so when all dials are set at 3 the dusting or printing of each of the three toners is equally dispersed on the paper's surface.

Also, these color adjustment dials allow for minor color corrections of full-color originals. Therefore, knowledge of color theory and its applications allows for fine tuning of each copy. For example, if an original is too green (examine the white areas and flesh tones for a greenish cast), the magenta must be increased, or the yellow and cyan can be decreased.

Marka Hitt worked with a Far Eastern theme to create the color Xerox shown above.

Susan Osborn's Self-Portrait began with a color Xerox. Osborn then used Prisma pencils, paint, and other conventional artist's media to add color and texture.

COLOR CORRECTING SLIDES

The color-contrast control and the color-adjustment dials have no real effect when color correcting slides. Because the slides are projected by a light source, they are corrected with photographic color correction filters (CC filters) between the lens and the ground glass. A trained Xerox 6500 Color Copy operator or technician can assist you in selecting the proper CC filters. The intensity of the light source can be increased or lowered, which has the same effect as the contrast-control dial in creating either rich, high-contrast color or more subdued, gradated renditions of a transparency's color.

FULL-COLOR SCANNING

With flat artwork, scanning can be done with the light source, just as with black-and-white machines. Each resulting copy is different, depending on when the motion occurs. For example, holding your subject still for the magenta and yellow passes but moving it with the light during the cyan pass creates a ghostlike image in shades from violet to cyan.

PAPER SELECTION

The Xerox Color Copier can accommodate many different types of paper. Standard 20 pound bond, either 8½ × 11 or 8½ × 14, is most often used in the Xerox 6500 machine. These lightweight papers need not be white. Interesting results can be achieved with colored stock. Heavier papers can also be used; a printer's card stock produces good results. Some people

To create this Xerox print, Susan Osborn cut a 4 x 5 Kodalith positive into sections the size of a 35mm slide. She then enlarged each section separately in a color Xerox machine to obtain a patchwork of different hues.

have even had success with such papers as Rives BFK or Arches.

Warning: You must check with the machine operator before inserting any papers other than the approved printing papers. Thicker papers can jam the machine or even cause fires.

The Xerox machine also prints on acetate sheets made by Xerox. The toners are dusted onto the plastic and then fused. The image must be hardened to prevent cracking and chipping. A Xerox Transparency Machine is supplied with all copiers for this purpose.

Xerox also supplies special transfer paper so that color copies can be transferred onto paper or fabric. This transfer paper is a latex-coated stock that accepts the toners into its plastic base. The resulting transfer is placed face down on either a larger sheet of paper or a piece of fabric and is transferred by the heat of either a hand iron at the linen setting or a dry-mount press set at 350° F. (177° C.) for 2 minutes.

It is important to note that this special application extends even further the image- and object-mak-

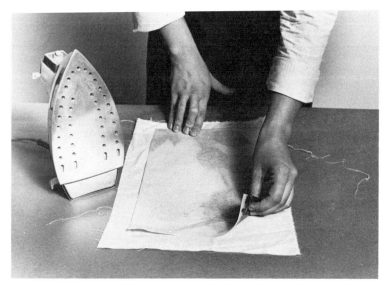

With the iron at the hottest setting, apply heat first in the center, and then work toward the edges.

Place the image copied onto the special transfer paper face down on the fabric's surface. Use natural fibers that can withstand high temperatures.

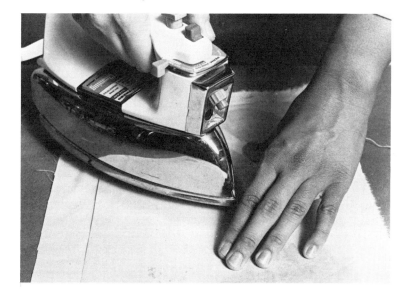

Extend the iron all the way out to the edges so that the transfer is complete from corner to corner and edge to edge.

ing capabilities of the Xerox 6500 Color Copier. Not only can larger collages be made with transfer paper, but full-color images can be added selectively as well as combined with the other nonsilver processes presented in this book.

As with the black-and-white machines, the creative applications of generation loss, overlays, sequencing, scanning, and the various methods of transferring (to paper or fabric) become viable selections in utilizing the Xerox 6500 Color Copier. The choices are yours to explore.

When the back of the transfer has received sufficient heat over its entire surface, quickly peel back the paper. Pull in a continuous motion or a line will show where you hesitated.

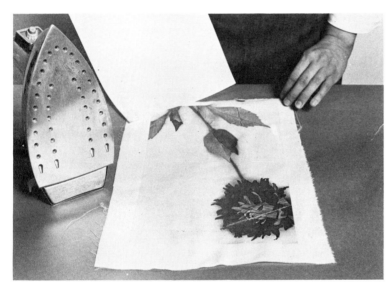

The transfer is now complete.

12

FINISHING YOUR IMAGES FOR PRESENTATION

Throughout this book you have been presented with many new processes, procedures, and techniques for making nonsilver and artistic photographic images. You have measured and mixed chemicals; selected, sized, and sensitized paper and fabric; developed images in everything from ammonia to soapsuds—and now you are finally confronted with still more choices to consider when finishing an image.

Toning, hand coloring, sewing, stuffing, and adding actual objects to the print are only a few of the infinite options available to you. Being aware of these visual possibilities is very important, because you must consider them from the beginning when previsualizing your initial idea. It is imperative to recognize all the possible choices and make the selection an integral part of the total image, not as an afterthought.

ADDING COLOR OVERALL

Toning is one of the most popular methods of adding overall color to a black-and-white silver print, and this is perhaps the best place to start. Toning is a procedure that alters the color of a black-and-white silver print. A chemical (or chemicals) in the toner reacts with the metallic silver, causing a new compound to form, thus creating a new color.

KODAK TONERS

Photographers such as Frederick Sommer and Ansel Adams use selenium toning not only to make blacks appear richer, but for archival permanence as well. Kodak's selenium toner, when used with an image printed on warm-toned paper, affects only the blacks without changing any of the other values of the print.

Kodak's sepia toner sufficiently changes the metallic silver to create an image in varying tonalities of sepia. This is a two-step process of a bleach bath (potassium ferricyanide diluted in water) and the sepia toner. Detailed instructions are included with the product that explain mixing as well as step-by-step guidelines for toning. Sepia toner gives off a foul odor, so a well-ventilated room is necessary. A hardening fixer must be used after toning to complete the process.

Sepia-toning a print prior to hand coloring with Marshall Oils has been the traditional procedure. It should be seriously considered if you wish to re-create photographs with a vintage look or to hand-color a modern image. Sepia toning warms the tone of the print and so prevents a harsh color relationship when the oils are added. Not only will sepia toner tone conventional silver prints a rich brown color, but it can also be tried on litho film and other silver materials.

HALO-CHROME

Halo-Chrome, manufactured by Rockland Colloid Corp., converts the black photographic image to pure metallic silver to render a silver-on-white print. It works with all paper surfaces (even resin-coated).

To use Halo-Chrome, immerse an already processed and completely washed print in the bleach bath supplied by the manufacturer until the bleaching action has stopped (approximately 3 to 5 minutes). Rinse for 1 minute, then immerse the print in a second bath of Halo-Chrome prepared just prior to this second step. Agitate the print until the desired silver tone is acquired or the Halo-Chrome's action stops (about 1 minute). Rinse, fix, and wash for 20 minutes. Air-dry it, emulsion side up. The final toned print must be protected from tarnishing, so cover it with plastic or acetate, or frame it behind glass. There is a way of reversing the tonal relationship so that the final image is black-on-silver, but it requires a darkroom, and only certain papers will work.

Caution: These chemicals are dangerous, so read all the instructions included in the package before you begin.

EDWAL COLOR TONERS

Sepia and silver are not the only colors available for toning a black-and-white print. Kodak manufactures blue and brown toners, and Edwal Scientific Products makes Edwal color toners in five colors: red, green, blue, brown, and yellow. These can be mixed to make other colors. **Note**: Edwal toners affect the entire surface of the print, coloring both highlights and shadows. If you wish white areas to remain as white as the base of the photo paper, it is imperative to mask these areas with rubber cement. This procedure is explained later in this chapter. No bleaching bath is required; just dilute the toner with water according to the package directions and immerse the print in the tray until the desired overall tone is achieved. No fixing is necessary; simply wash the print for 20 minutes. These are not very stable toners, and they will fade and alter significantly if placed in direct sunlight.

OTHER CREATIVE ASPECTS

Toning is not limited to black-and-white photographs. Other surfaces and materials accept toners to varying degrees. Selectively masking out areas prior to toning and using paintbrushes on smaller areas allow for creative experimentation.

The Luminos Pastel Photo Papers (described in Chapter 9) accept toners well; the finished effect is a toned image on a pastel background. The harsh blacks are warmed by sepia toning. A monochrome image can be created by using Edwal blue toner on pastel blue paper.

MASKING OUT AREAS

Selective toning is accomplished by using rubber cement as a liquid-resisting mask. Brush on a generous amount of rubber cement, applying it to an area you *do not wish to color*. (If it is too thick to spread evenly, thin with rubber cement thinner at about a 1:2 ratio.) Use a smaller paintbrush when masking out small areas and for along edges. Immerse print in the toning solution following the toner's specific instructions. When the print is completely dry, remove the rubber cement using the tips of your fingers, picking it up in small balls.

You can then coat the toned areas with rubber cement, and tone the uncovered areas in another color.

Edwal toners are translucent. With selective masking and repeated tonings with different colors rich color combinations will build up. Layering the color in this manner allows more control of the toning process. Results achieved with multiple toning are often more interesting than those accomplished by simply mixing two toners.

TONING DETAILED AREAS

Small detailed areas in a photograph can be toned using paintbrushes or swabs. Make sure the paintbrushes do not have metal edges around the top of the bristles, because most toners react chemically with metallic substances. Purchase plastic-handled brushes, or use the oriental-style watercolor brushes that have bamboo shafts. Follow the package directions for each specific toner. Then, instead of immersing the entire print in a tray, place it on a flat, hard surface and apply toner with the brush in only the areas you wish to color. Combining and layering colors and simultaneous multiple toning of different areas can be done. Wetting the print using a sponge with water and then applying toner to the damp surface creates the watercolor technique of wet-on-wet (the toner spreads out like a puddle, making a fluid line of color). Use rubber cement to protect specific areas.

The possibilities go on and on, and it is up to you to pursue these options further. Just remember that the toners are only reacting with the silver image of the print. To color the highlights and the unexposed paper surface, dyes or other paper-staining solutions must be used.

DYES

Coloring the highlights, unexposed areas, and the entire print's surface of both silver image and paper fiber is accomplished by the use of dyes and other colorants.

Rockland Colloid Corp.'s Prin-Tint is a concentrated colorant for use with black-and-white prints (fiber-base and resin-coated) and for coloring images on artist's paper such as blueprints or brownprints. Red, yellow, and blue primary colors, plus a blending concentrate, are included in a box of PrinTint, which has the capacity to tint up to 300 8 × 10 prints. Rubber cement can be used as a mask while coloring other areas. These

dyes are recommended for retouching color photographs and transparencies, too.

Dr. Martin's Dyes are artist's dyes available in most art supply stores. They are highly concentrated dyes that work well on any artist's paper when diluted with water. They can also be used to dye black-and-white fiber-base prints. Wet-on-wet and other watercolor techniques can be utilized with Dr. Martin's Dyes.

Another excellent product is Marhsall's Photo Retouch colors, manufactured by the John G. Marshall Co. These are water-base concentrated dyes that can be used for coloring an entire print or for retouching. Unlike PrinTint and Dr. Martin's Dyes, these can be removed if applied too heavily.

To remove, simply dilute one drop of household bleach to ten drops of water. Apply this to the dye and allow it to sit until the color is the desired density or is completely removed. To stop the bleaching action, blot the solution with cotton and flush the surface with water. Wash the print completely before a water rim or an edge occurs from uneven application. Reapply additional color or a new color.

Finally, do not forget to experiment with commonly available dyes such as Rit and household food colorings. Both have the ability to stain papers or, for that matter, fabrics.

Combining toners with dyes is another option. Dyeing a print and then toning only the silver image sets up unique color relationships. For example, using PrinTint first and following it with Halo-Chrome creates a metallic photographic image on a colored background. Other techniques include using

rubber cement as a mask, detailing with a paintbrush or cotton swab, and using either or both solutions and manipulating or combining them.

HAND COLORING

There are many ways to hand color both silver and nonsilver photographs. Product selection depends on the surface and the image itself. Experimentation and some experience in working with hand-coloring materials helps also.

MARSHALL OILS AND PENCILS

Marshall photo oils are the best-known colors for hand coloring black-and-white prints. The paints are transparent, allowing the details of the photographic image to show through. Oils take a considerable amount of time to harden and dry, so coloring with them need not be finished in one session. Prints can be reworked until the desired results are obtained.

These oils are available in individual tubes, or they can be ordered as a kit. There are six different sets, ranging from the Introductory set of five tube colors to the Master set of forty-six oil paints. They are available at most camera, art, or graphic arts stores.

Marshall Oils can be applied to the following surfaces: nonglossy fiber-base photographic papers, matte-finish RC prints, negatives, and canvas. With the prior application of Marshall's Pre-Color spray, a toothing lacquer that causes oils to adhere, the following surfaces can be colored: glossy fiber-base and RC photos, Polaroid Land and SX-70 prints, and newspaper and magazine images.

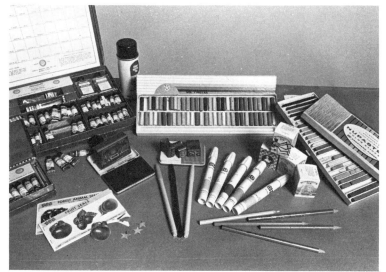

Colored pencils, pastels, markers, rubber stamps, decals, and Marshall's photo oil colors are just a few of the artist's supplies that can be applied to an image.

APPLYING MARSHALL OILS

"No knowledge of drawing or art training is . . . necessary." This line from the instruction book for the Marshall Oils is a bit misleading. The hand-coloring process is relatively simple, but some basic guidelines are worth noting, and some art training is of course always helpful.

Detailed instructions are supplied, and helpful step-by-step methods are included. For example, "How to Color a Landscape in Easy Stages" directs you on how to paint sky, clouds, and trees. Moreover, it explains which colors to use and which areas to work on first.

Simply stated, this is how to color a photographic image with Marshall Oils:

1. Prepare a workable palette of colors by placing a piece of clean glass over the color chart provided. Squirt out enough paint from each tube onto the glass next to the color's name. These oils are concentrated, so very small amounts will do. Have plenty of cotton or cosmetic swabs on hand with which to apply the paint.

2. If a glossy surface is to be colored, spray the print with Marshall's Pre-Color spray. Apply it first

Place clean glass over the printed palette provided by the manufacturer. Squeeze out enough oil paint for your painting session. Extender may be used to make a color more transparent.

in one direction. Let it dry, and reapply it in the opposite direction. Nonglossy surfaces need not be prepared unless the surface grips the color too strongly and the oils become hard to blend. Apply a light coat of Prepared Medium Solution (P.M.S.) to correct this problem.

3. Pick up a small amount of the desired color on the tip of a cotton swab or a cotton-tipped skewer, and apply it to the area of the print you wish to color. Rub the swab around, working the color in. To paint large areas, use a cotton ball to apply the paint (see illustration).

Apply paint with a cotton swab, working it into the image's surface.

Color large areas using a cotton ball, or blend with the cotton ball to achieve subtle areas of color.

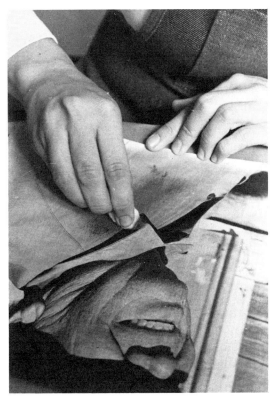

4. To pick up excess oil paint, use a clean swab and wipe away the color.

5. If you want a more subtle hue, dilute the color with the clear Extender, either mixing it with the paint on the palette or adding it on top of the already painted surface, blending it into the existing color.

6. Always rub down each colored area and remove excess color before painting an adjacent area.

7. You can completely remove a color from the print by using the Marlene solution. Put a small amount of solution on a swab, and rub the area, blotting and drying with clean cotton until the area is free of color. (*Note:* Marlene *cannot* be used with prints treated with Pre-Color spray, because this solution removes the toothing lacquer.)

8. Once the oil-painted print has dried, spray it with a fixative to protect the painted surface. Spray Goss is a glossy fixative and Pro-Tek-To Spray is a matte spray that eliminates glare and surface reflection.

Working with these oils can be quite frustrating at first, but be patient and keep at it.

Marshall also makes Photo-Painting pencils, which simplify the oil-painting process and eliminate some of the mess and odor of the paints. Marshall has discontinued making sets of the photo pencils, but individual pencils are available. If you purchase individual pencils from Marshall, request an instruction sheet.

Marshall Oils can be applied to Luminos Photo Linen with excellent results. Nonsilver prints on sized artist's paper accept the oil paints to varying degrees, but it is wise to use the Pre-Color spray to avoid any staining caused by the oil binder in the paints. Liquid Light, because it has a colloid binder, can be colored by the Marshall products, and of course pastel photo papers work, too. Try the oils on everything and record your results for use with later images.

OTHER HAND-COLORING MATERIALS AND METHODS

In addition to the Marshall products, there are several other materials and hand-coloring methods for photographic images. They include: Peerless transparent watercolors for photos, watercolors such as Winsor & Newton, india inks, color pencils, oil pastels, crayons and chalk pastels, alcohol markers and sprays, and spray paints.

Peerless transparent watercolors are manufactured exclusively for application on black-and-white fiber-base silver prints. They come in blotter form in a booklet of fifteen colors on paper. Just wet the brush, touch it to the blotter color, and apply it to the print's surface. The color is immediately absorbed into the paper, which makes mistakes difficult to correct. Peerless watercolors are harder to work with than Marshall oils, but the results are more subtle.

These watercolors can be tried on nonsilver images as well, but it is best not to limit yourself to a fifteen-color palette. Because nonsilver images are printed on artist's papers and various fabrics, you can experiment with ordinary watercolors. Brands such as Winsor & Newton, Grumbacher, and Permanent Pigments, used for gum printing and Kwik-Printing, are also excellent for hand coloring. Winsor & Newton also manufactures metallic water-base colors—silver and gold—that are opaque when applied.

India inks and other brands of opaque liquid colors work well on nonsilver images. These are absolutely permanent and cannot be removed, so carefully consider where they are to be applied.

A workable matte fixative spray must be applied to glossy surfaces before pencils, pastels, crayons, chalks, markers, and sprays can be used on the print's surface. Marshall's Pre-Color spray is excellent, but any artist's workable fixative will do.

Any nonsilver image printed on sized or unsized artist's paper will accept all of the above with excellent but varying results. Kwik-Prints do have a toothed surface, but a workable fixative may be needed. Experiment on the back of the print to test each form of hand coloring. Alcohol markers affect the toners of color Xeroxes, but this bleeding or running of the copy's colors is very interesting and resembles the wet-on-wet watercolor technique.

Not only can toning or dyeing be tried on fabrics, but selective hand coloring can also be done, depending on the weave of the material. Certain methods work better than others. Satins, finely woven silks, and smooth cottons accept colored pencils, oil pastels, chalks, and alcohol markers quite easily. Wider-weave fabrics accept alcohol markers unevenly and create a blotchy pattern of color. Using spray alcohol markers is preferable for correcting this. Color Xerox transfers on fabric accept all of these colors, except the alcohol markers. Sprays cause the toners to bleed and stain the fabric's edges. All of these, however, are techniques worth investigating.

Hand coloring is a very personal method for adding color to a photographic image. There are several how-to hand-coloring books on the market that can help you see how others have used the technique. At first, it is best to go with what feels most comfortable, working up areas slowly until you gain confidence with the different

This brownprint on artist's paper readily accepts hand coloring.

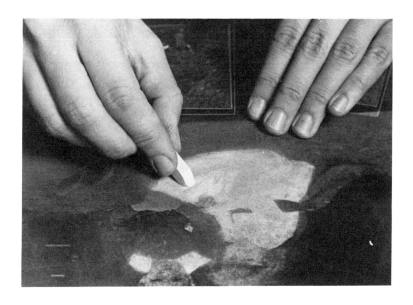

Here the artist uses colored chalk to create texture and mood in a brown-print on paper.

Stick decals, tape, stars, and other adhesive-backed objects onto the surface of an image.

materials and their results. Then, challenge yourself! Previsualize the image, including any, or all, of the finishing methods discussed, then work with them until your requirements are met.

FINISHING TECHNIQUES FOR FABRICS

Working with images on fabrics sets up entirely different finishing problems from those encountered when hand coloring a photograph and mounting it in a frame. Fabrics can be molded, shaped, stuffed, distorted, or otherwise physically manipulated. One of the inherent characteristics of fabric is its flexibility; it accepts both stretching and stuffing reasonably well. Stretching a finished photographic image around stretcher bars or a wooden frame is an obvious finishing technique, but it does not showcase the full dexterity of the fabric, nor does it challenge the dimensionality of the photographic image.

Slight stuffing and gross exaggeration of the physical image—or anywhere in between—are choices you may consider. The procedure for stuffing is called *trapunto*. The procedure is simple:
1. Place the finished fabric over another piece of fabric. Any type of fabric will do for the backing, so stretchable synthetics can be used with natural fibers and vice versa.
2. Baste the two pieces together by hand, using large stitches.
3. Machine-stitch the photographic image and the backing together, outlining areas with threads

the same color as the image or with contrasting colors. Close up specific shapes, isolating areas to be stuffed later.

4. When the top stitching is complete, turn the fabric over and cut or slit the backing only, being very careful not to pierce the front piece of the finished fabric.

5. Stuff the space with quilting batting until the desired depth is attained. Hand-sew the opening closed, pulling the edges together so that no batting will fall out.

6. Repeat the procedure for other areas.

Like the Photo Linen illustrated here, many images can be sewn, stuffed, and otherwise manipulated to enhance the idea behind the image.

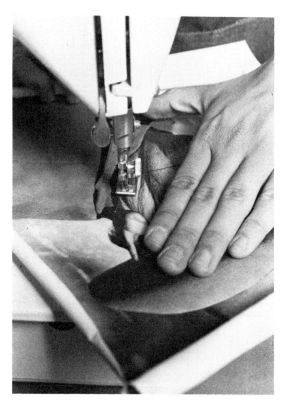

Trapunto can be done to varying degrees, creating multiple layers in the image's surface. Stuffing can even be allowed to fall out from the image itself.

Machine stitching with matching or contrasting threads is similar to working with pencils or crayons on the surface of a paper print. Hand embroidery, using different thicknesses of threads and various stitches, is worth researching. There are several sewing and quilting books now available in fabric

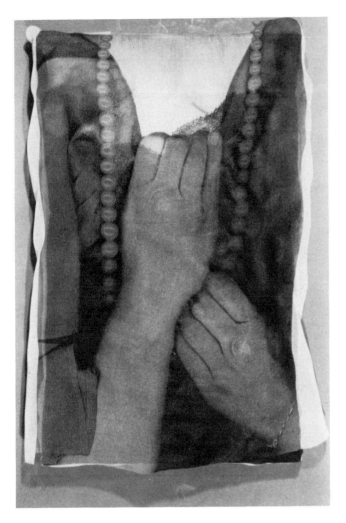

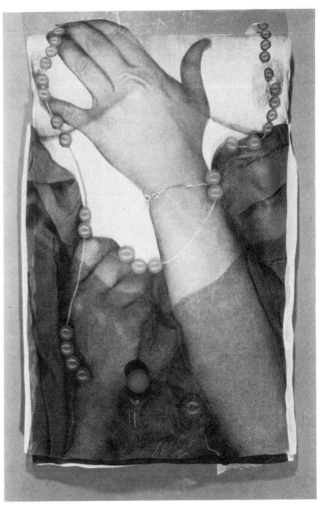

The three images shown above are from a sequence of seven entitled She Felt Condemned by Her Own Femininity. *Suda House began her work by transferring color Xeroxes to chiffon and satin. She then stitched and stuffed the pieces and added color with Prisma pencils. The last two pieces in the series include real beads.*

stores that showcase all of these sewing crafts.

In terms of presentation, you should consider whether to finish the edges of the fabric piece. As previously mentioned, one choice is to stretch the fabric around stretcher bars or a wooden frame. Hemming or binding the edges to another backing fabric is also an excellent finishing method. But do not rule out leaving the edges unfinished. Often the rough, raw edges of the fabric, with the thread left hanging, work well with the content of the image.

Incorporate some of the paraphernalia associated with sewing. Needles, pins, buttons, zippers, Velcro, embroidery hoops, and batting can be juxtaposed with printed images. Working photographically with fabrics and solving their special presentation problems allows the nonsilver and artistic photographer to create a three-dimensional, photographic illusion.

I have repeatedly emphasized that the possibilities are endless, but it is a statement of fact. This book could go on and on to explore every subtle nuance of nonsilver and artistic photography, but I've left blank pages at the end for note-taking, so that you can continue in the direction *you* want. Push your ideas, challenge the processes, and break the rules!

Notes

APPENDIX A | How to Make a Halftone Screen

Making a Halftone Screen

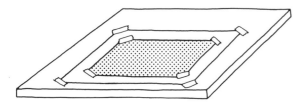

Transfer side face down
on acetate sheet

Brayer used to
apply pressure

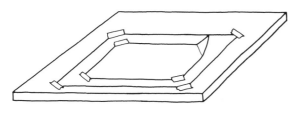

Area shown pulled
back or loosened

Supplies:

- A clear sheet of acetate or plastic, available at most art supply stores

- A full sheet of press-transfer dots, lines, or other patterns, available at most art supply stores and manufactured under such brand names as Zipatone, Let-R-Set, or Chartpak

- A hard rubber brayer or a metal spoon

Procedure:

1. Cut the acetate sheet so that it is slightly larger than the litho film size you will be using.

2. Tape the acetate sheet down on a tabletop or other hard surface. Some plastic sheets have a rough side the same as a photo emulsion. This rough side, if apparent, should be face up.

3. Tape the press-transfer patterned sheet over the acetate. Secure all corners.

4. Apply pressure with the brayer to the center of the sheet and work out toward each corner.

5. Be careful to apply even pressure, and use your hand to hold the press-transfer sheet in contact with the acetate. If necessary, loosen the taped edges one corner at a time, to avoid cracking the transfer sheet as you rub.

6. Transferring is complete when the press-transfer sheet is frosty in appearance. *Note:* Rub any areas that do not stick with a metal spoon. Apply pressure to the loose areas only.

7. Protect the homemade halftone screen with a bond paper slip sheet and store it flat in a cool, dry place.

How to build a light box | APPENDIX B

Supplies:

- Start with a wooden box of ½- or ¾-in. (13 or 19 mm) plywood or particle board construction. An old wooden drawer with even edges is a good choice. The size of the box (width and length) should be determined by the largest size image you will wish to print. For example, if you plan to print 11 × 14, the box's dimensions should be slightly larger. The depth of the box depends on the number of bulbs, their placement, and the factors of the inverse square law which will be explained later. It is best to construct the box with deep sides, which can be trimmed down later.

- A sheet of heavy-duty, double-thick safety glass with smoothed, sanded edges to fit on top of the box. *Note:* A groove can be cut and the glass set in and secured to avoid slippage.

- The following electrical parts:
  a. Fluorescent, unfiltered ultraviolet (black light) bulbs. The length of the bulbs are determined by the size of the box.
  b. Ballast units: There are two types. The conventional models which require starters and the self-starting models. Self-starting ballasts require rapid-start bulbs, which for unfiltered ultraviolet are large. The smaller bulbs require conventional ballasts with starters.
  Note: Every two bulbs requires a ballast.
  c. Two fluorescent "tombstones" or end sockets for each fluorescent tube.
  d. A starter and starter socket for each bulb if a conventional ballast is used.
  e. Electrical wiring sufficient to create a circuitry system, between the bulbs, the starters, and the ballasts. Optional: A switch to turn the current off and on. A double-two conductor or three conductor with ground cord with plug for inserting in electrical wall socket is required.

Procedure:

1. After constructing a box or finding one, if you are using self-starting ballasts the box must be lined with aluminum foil.

2. Spacing of the bulbs is important. If they are too close, bands of overexposure will appear on the image. If bulbs are placed too far apart, the exposure will be uneven from edge to edge on the print. The distance of the bulbs from the top of the box must also be considered. Bulbs placed far apart and too far away from the glass top will create long exposure times. The best arrangement is to construct the bulb arrangement so they are evenly spaced along the entire bottom of the box.

3. Place the tombstones (a pair for each bulb) opposite one another so that the distance between them equals the length of the bulb. Insert a bulb in each pair of tombstones, turning them and locking all three pieces together. Do this to all the bulbs. Space them evenly as discussed above, leaving room for the ballasts or starters. Although these items can be placed outside the bulb arrangement, this will require the use of more electrical wiring. Tape down the end pieces with drafting or masking tape.

4. Place and tape down the ballasts or starter sockets according to the diagram on the label of the ballast. A starter unit should be spliced between the ballast and each individual bulb.

5. Connect all electrical parts with wiring, tying off wires and using electrical tape if necessary (see diagram). **Note: Discussing the wiring of this entire unit with a trained and qualified electrician or electrical hardware salesman is advisable.**

6. Remove the bulbs and screw in the tombstones, the ballasts or the starter sockets.

7. Drill a hole in one side near the bottom of the box. Splice the two- or three-conductor cord with ground to the interior wiring. Splice a wall plug onto the exterior wire (two- or three-conductor cord). An On/Off switch is optional.

8. Use this light box in a dimly lit or light-safe area when handling the coated materials. Set the transparency on the glass top. Place the sensitized paper or fabric directly on top of the film. Then place a heavy, flat board on top so that 100 percent contact is created. Cover the light box with an opaque cover such as thick black plastic, or a dark cloth. The light given off by the unfiltered ultraviolet could damage your eyes. DO NOT LOOK DIRECTLY AT THE BULBS.

9. This is a consistent ultraviolet light source. Calculate exposure times for each process and negative type. Record this information so exposures can be repeated without depending on the sun.

How to build a
light box

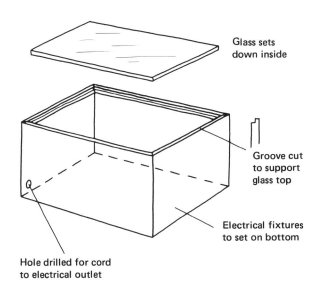

Glass sets
down inside

Groove cut
to support
glass top

Electrical fixtures
to set on bottom

Hole drilled for cord
to electrical outlet

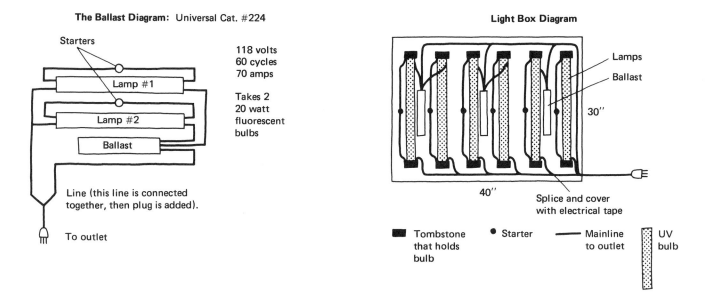

The Ballast Diagram: Universal Cat. #224

Starters

Lamp #1

Lamp #2

Ballast

118 volts
60 cycles
70 amps

Takes 2
20 watt
fluorescent
bulbs

Line (this line is connected
together, then plug is added).

To outlet

Light Box Diagram

Lamps

Ballast

30″

40″

Splice and cover
with electrical tape

Tombstone
that holds
bulb

Starter

Mainline
to outlet

UV
bulb

How to Build a Roller Squeegee | APPENDIX C

Roller Squeegee

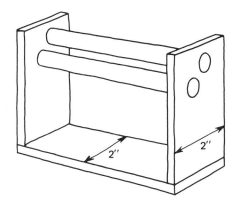

Side View to Drill Holes

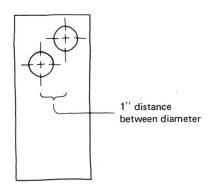

1'' distance between diameter

Direction to pull paper through squeegee

Photo tray

Supplies:

(Measurements are for use with 11 × 14 paper)

- A 4-ft. (1.2-m) length of 1-×2-in. (25-×51-mm) pine

- Two wooden dowels ½ to ¾ in. (13 to 19 mm) in diameter, each 18 in. (46 cm) long

- Finishing nails, glue, medium-weight sandpaper, a drill, and a hammer

- Polyurethane, resin-based coating, or other water-resistant paint.

Procedure:

1. Cut the 1-×2-in. pine into three lengths, one measuring 16 in. (38 cm) and two measuring 12 in. (30 cm). These are approximate measurements and can be increased so that larger papers can be squeegeed later.

2. On the 2-in. side of the 12-in. lengths drill two holes for the dowels. They must be staggered, one higher than the other, 1 in. (25 mm) apart, measuring from the center of the circle.

3. Nail the two 12-in. lengths upright at each end of the 18-in. length, creating a U-shaped stand.

4. Insert the dowels through the drilled holes and nail or glue them in place. With sandpaper smooth out all rough edges, especially on the dowels.

5. Waterproof the entire apparatus using any water-resistant paint.

6. To use the apparatus for sizing papers, simply place the roller squeegee in the bottom of a photo tray so that sizing can be collected and recycled. After soaking paper in sizing solution, drain it. Insert the saturated paper between the dowels, so that the lower one is underneath and the upper one is above the paper (see diagram). Pull the paper through, dragging the top of the paper against the top dowel and the back of the paper against the bottom one—creating an S shape for the squeegeeing motion.

7. Hang the paper to dry. Rinse off the squeegee and wipe it dry for later use.

APPENDIX D | Sources of Supply

A. Daigger & Co.
157 W. Kinzie St.
Chicago, IL 60610
Chemicals and scientific supplies.

Air Photo Supply Corp.
33 Wolfe St.
Yonkers, NY 10705
Luminos Photo Linen and pastel papers.

Berg Color Tone
Dept. D
P.O. Box 16
East Amherst, NY 14051
Toners. Write for information.

Curtin Matheson Scientific Inc.
1103 Slocum St.
Dallas, TX 75222
Chemicals and scientific supplies.

D.Y.E. Textile Resources
3763 Durango Ave.
Los Angeles, CA 90034
Pre-measured blueprint and brownprint kits.

Edwal Scientific Products Corp.
12120 S. Peoria St.
Chicago, IL 60643
Edwal color toners.

Fisher Scientific Co.
2225 Martin Ave.
San Francisco, CA 95050

7464 Chancellor Dr.
Orlando, FL 32809

1600 W. Glenlake Rd.
Chicago, IL 60143

10700 Rockley Rd.
Houston, TX 77001
Chemicals and scientific supplies.

Free Style Sales Co., Inc.
5124 Sunset Blvd.

Los Angeles, CA 90027
Mail-order house for a variety of litho films, pastel photo papers, some Rockland Products, and Premier Contact Frames.

ICN Life Sciences Group
121 Express St.
Plainview, NY 11803

2727 Campus Dr.
Irvine, CA 92664

26201 Miles Rd.
Cleveland, OH 44128
Chemicals and scientific supplies.

John G. Marshall Mfg. Co.
P.O. Box 649
Deerfield, IL 60015
Marshall oil paints, photographic pencils, retouching colors, and assorted hand-coloring supplies.

Light Impressions Corp.
131 Gould St.
Rochester, NY 14610
Exclusive distributor for Kwik-Print materials and instructional set of eighty slides with accompanying 20-minute tape. Light Impressions also distributes a wide variety of books. Ask for a catalog.

Luminos Photographic
25 Wolfe St.
Yonkers, NY 10705
Complete line of all Luminos photographic products.

McManus & Morgan
2506 W. Seventh St.
Los Angeles, CA 90005
Artist's paper, transfer and lift materials, and a variety of hand-coloring supplies.

Naz-Dar Co.
1087 N. Branch St.
Chicago, IL 60622

2832 S. Alameda St.
Los Angeles, CA 90015
Silk-screen extender base no. 5536 for magazine transfers.

Peerless Color Laboratories
11 Diamond Place
Rochester, NY 14609
Transparent watercolors for photographs.

Photographer's Formulary, Inc.
P.O. Box 5105
Missoula, MT 59806
Chemicals supplies for many non-silver processes, some prepared kits. Write for catalog.

Rockland Colloid Corp.
302 Piermont Ave.
Piermont, NY 10968
Complete line of Rockland products—Liquid Light, Fabric Sensitizer FA-1, Halo-Chrome, PrinTint—and others. Write for New Horizons in Photography, *a catalog of Rockland products.*

Salis International
4093 N. 28th Way
Hollywood, FL 33020
Dr. P.H. Martin's Transparent Watercolors. Ask for catalog and list of distributors.

Sangray Corp.
P.O. Box 2388
Pueblo, CO 81004
Distributes Decalon, an instant decal/lift medium. Booklet Decal'comania, The Art of Making Decals *(1975) is available.*

Screen Process Supply Co.
1199 E. Twelfth St.
Oakland, CA 94606
Inkodyes.

FURTHER READING

AA-5/*How to Make and Use a Pinhole Camera*, 9-78. Rochester, N.Y.; Eastman Kodak Co.

AA-9/*A Glossary of Photographic Terms*, 7-79. Rochester, NY; Eastman Kodak Co.

AG-2/*Photograms—Photography Without a Camera*, 10-77. Rochester, NY; Eastman Kodak Co.

AJ-5/*Photographic Sensitizer for Cloth and Paper*, 7-77. Rochester, NY; Eastman Kodak Co.

The Art of Photography. Life Library of Photography series. New York: Time-Life Books, 1971.

Boyd, Harry, Jr. *A Creative Approach to Controlling Photography*. Austin, TX: Heidelberg Publishers, Inc., 1974.

Brooks, David. *How to Select and Use Photographic Materials and Processes*. Tucson, AZ: HP Books, 1979.

Burchfield, Jerry. *Darkroom Art*. New York: Amphoto, 1981.

Cavallo, Robert M., and Kahan, Stuart. *Photography: What's the Law?* New York: Crown Publishers, Inc., 1979.

Coke, Van Deren. *The Painter and the Photograph; From Delacroix to Warhol*, rev. ed. Albuquerque, NM University of New Mexico Press, 1972.

Color. Life Library of Photography series. New York: Time-Life Books, 1971.

Crawford, William. *The Keepers of Light: A History and Working Guide to Early Photographic Processes*. Dobbs Ferry, NY: Morgan & Morgan, 1979.

Creative Darkroom Techniques. Rochester, NY: Eastman Kodak Co. 1975.

Davies, Thomas L. *Shoots: A Guide to Your Family's Photographic Heritage*. Danbury, NH: Addison House, 1977.

Dolloff, Francis W., and Perkinson, Roy L. *How to Care for Works of Art on Paper*. Boston: Boston Museum of Fine Arts, 1971.

Electroworks. Introduction by Marilyn McCray. Rochester, NY: catalog from the International Museum of Photography, George Eastman House, 1979.

Firpo, P.; Alexander L.; Katayanagi, C.; and Ditlea, S. *Copy Art*. New York: Richard Marek Publishers, 1978.

Fraprie, Frank R., and Woodbury, Walter E. *Photographic Amusements*, 10th ed., reprint of 1931 ed. New York: Literature of Photography Series, Arno Press.

Frobisch, Dieter, and Lamprecht, Hartmut. *Graphic Photo Design*. New York: Amphoto, 1978.

GA-11-4/*Copy Preparation*. Rochester, NY: Eastman Kodak Co.

Gassan, Arnold H. *Handbook for Contemporary Photography*, 4th ed. Rochester, NY: Light Impressions, 1977.

Gill, Arthur T. *Photographic Processes*. Rochester, NY: International Museum of Photography, George Eastman House.

Golwyn, Craig. "This Issue: Electrostatics, State of the Science, State of the Art, State of Their Union." *Yony* 1, no. 4 (May 1975). Art Institute of Chicago.

Green, Jonathan, ed. *Camera Work: A Critical Anthology*. Millerton, NY: Aperture, 1975.

Grimm, Tom. *The Basic Darkroom Book: A Complete Guide to Processing and Printing Color and Black-and-White Photographs*. New York: New American Library, 1978.

Gross, Henry. *Simplified Bookbinding*. New York: Charles Scribner's Sons, 1976.

Howell-Koehler, Nancy. *Photo Art Processes*. Worcester, MA: Davis Publications, Inc., 1980.

Hunt, Robert W. *Reproduction of Colour in Photography, Printing, and Television*, 3d ed. New York: Halstead Press, 1976.

J-1/*Processing Chemicals and Formulas*, 1-77. Rochester, NY: Eastman Kodak Co.

Kosar, Jaromir. *Light-Sensitive Systems: Chemistry and Application of Nonsilver Halide Photographic Processes*. New York: John Wiley & Sons, 1965.

Land-Weber, Ellen. "3M Montage to Create Dreamlike Images Using Modern Technology." *Petersen's PhotoGraphic Magazine*, May 1976, pp. 72–76.

Lewis, Arthur W. *Basic Bookbinding*. New York: Dover, 1952.

Lietze, Ernest, and Lyons, Nathan, eds. *Modern Heliographic Processes: A Manual of Instruction*. Rochester, NY: Visual Studies Reprint Series, 1974.

Lundquist, Par. *Photographics: Line and Contrast Methods*. New York: Van Nostrand Reinhold Co., 1972.

Margolis, Marianne F., and Phillips, Christopher. *Steichen: A Centennial Tribute*. Rochester, NY: International Museum of Photography, George Eastman House, 1979.

Marino, T. J. *Pictures Without a Camera*. New York; Sterling Publishing Co., 1975.

Neblette, C. B. *Photography, Its Principles and Practices*, 2d ed. New York: D. Van Nostrand Co., 1930, pp. 481–93.

Nettles, Bea. *Breaking the Rules*. Rochester, NY: Inky Press, distributed by Light Impressions, 1977.

Parker, Fred R., ed. *Attitudes: Photography in the Seventies*. Santa Barbara, CA: Santa Barbara Museum of Art, 1979.

Petersen's Guide to Creative Darkroom Techniques. Los Angeles: Petersen Publishing Co., 1973.

Podracky, John. *Photographic Retouching and Air-Brush Techniques*. New York: Light Impressions, 1980.

Q-1/ *Basic Photography for the Graphic Arts*, 10-77. Rochester, NY: Eastman Kodak Co.

Q-2/ *Kodak Photographic Materials for the Graphic Arts*, 9-76. Rochester, NY: Eastman Kodak Co.

Q-3/ *Halftone Methods for the Graphic*, 8-78. Rochester, NY: Eastman Kodak Co.

Q-7/ *Basic Color for the Graphic Arts*, 7-77. Rochester, NY: Eastman Kodak Co.

Routh, Robert D. *Photographics*. Los Angeles: Petersen Publishing Co., 1976.

Scopick, David. *The Gum Bichromate Book: Contemporary Methods for Photographic Printing*. Rochester, NY: Light Impressions, 1978.

Shull, Jim. *The Hole Thing: A Manual of Pinhole Fotography*. Dobbs Ferry, NY: Morgan & Morgan, 1974.

Swedlund, Charles. *Photography: A Handbook of History, Materials, and Processes*. New York: Holt, Rinehart & Winston, 1974.

Towler, John. *The Silver Sunbeam*, reprint of 1864 ed. Dobbs Ferry, NY: Morgan & Morgan, 1969.

U-750/ *Graphic Arts Literature Packet*. Rochester, NY: Eastman Kodak Co.

Vogel, Herman. *The Chemistry of Light and Photography*. New York: Arno Press, 1973.

Wade, Kent E. *Alternative Photographic Processes*. Dobbs Ferry, NY: Morgan & Morgan, 1978.

Walker, Sandy, and Rainwater, Clarence. *Solarization*. New York: Amphoto, 1974.

Wall, Alfred H. *A Manual of Artistic Colouring, As Applied to Photographs*, reprint of 1861 ed. New York: Literature of Photography Series, Arno Press, 1973.

Wall, Edward John. *Photographic Facts and Formulas*, revised by Franklin I. Jordan. Boston: American Photographic Publishing Co., 1974.

Weinstein, Robert A., and Booth, Larry. *Collection, Use, and Care of Historical Photographs*. Nashville, TN: American Association for State and Local History, 1977. Distributed by Morgan & Morgan.

Wheeler, Owen. *Photographic Printing Processes*. London: Chapman & Hall, Ltd., 1930.

Wilhelm, Henry. *Preservation of Contemporary Photographic Materials*, Grinnell, IA: East Street Gallery, 1975.

Zakia, Richard D. *Perception and Photography*. Rochester, NY: Light Impressions, 1979.

Zakia, Richard D., and Todd, Hollis N. *Color Primer I & II*. Dobbs Ferry, NY: Morgan & Morgan, 1974.

——— and ———. *101 Experiments in Photography*. Dobbs Ferry, NY: Morgan & Morgan, 1969.

Index